Digital Earth
Cyber threats, privacy and ethics in an
age of paranoia

Digital Earth
Cyber threats, privacy and ethics in an age of paranoia

SARAH KATZ

IT Governance Publishing

Every possible effort has been made to ensure that the information contained in this book is accurate at the time of going to press, and the publisher and the author cannot accept responsibility for any errors or omissions, however caused. Any opinions expressed in this book are those of the author, not the publisher. Websites identified are for reference only, not endorsement, and any website visits are at the reader's own risk. No responsibility for loss or damage occasioned to any person acting, or refraining from action, as a result of the material in this publication can be accepted by the publisher or the author.

Apart from any fair dealing for the purposes of research or private study, or criticism or review, as permitted under the Copyright, Designs and Patents Act 1988, this publication may only be reproduced, stored or transmitted, in any form, or by any means, with the prior permission in writing of the publisher or, in the case of reprographic reproduction, in accordance with the terms of licences issued by the Copyright Licensing Agency. Enquiries concerning reproduction outside those terms should be sent to the publisher at the following address:

IT Governance Publishing Ltd
Unit 3, Clive Court
Bartholomew's Walk
Cambridgeshire Business Park
Ely, Cambridgeshire
CB7 4EA
United Kingdom
www.itgovernancepublishing.co.uk

First published in the United Kingdom in 2022 by IT Governance Publishing.

ISBN 978-1-78778-359-1

ABOUT THE AUTHOR

Sarah Katz is a graduate of UC Berkeley with a background in cybersecurity. She works as a technical writer at Microsoft®, and previously worked at NASA. Her writing has appeared in *Cyber Defense Magazine*, *Dark Reading*, *Infosecurity Magazine*, and *Tech Xplore*. Sarah has served as a cybersecurity speaker for the Brazilian technical institute IGTI, and is pursuing a PhD in cyberpsychology with a focus on user security in augmented reality. The short film *Rationale* – currently in production – based on her eponymous short story and published in *Thriller Magazine*, is about the risks of using mood-altering technology to enhance cognition.

ACKNOWLEDGEMENTS

I would like to thank Allen Joe who was my supervisor during my time at NASA, and who served as a main source of inspiration for this book.

I would also like to thank Yinka Akingbehin, Chris Evans, and Christopher Wright for reviewing this book during the production process.

CONTENTS

Contents

INTRODUCTION

Information security concerns such as foreign meddling in politics, such as the 2016 US presidential election[1] and phishing campaigns that used the COVID-19 pandemic to intimidate victims into engaging with malicious email,[2] have made user security awareness a chief priority for the cybersecurity space. Whether it's news of the next big ransomware attack or attackers turning to the Internet to commit terrorism, a multitude of threats have evolved. From phishing to political interference to cyber terrorism to police surveillance, issues of data and physical security as well as privacy continue to arise in the use of computers for both the individual user and large enterprise alike.

In fact, as many threat actors seek to attack the government infrastructure of powerful nations as well as the personal information of citizens, the United States federal and private health care sectors as well as users of social media and wireless devices have all become significant target risk

---

[1] "2016 Presidential Campaign Hacking Fast Facts." *CNN*, October 28, 2020, *https://edition.cnn.com/2016/12/26/us/2016-presidential-campaign-hacking-fast-facts/index.html*.

[2] Brumley, K. "COVID-19 Scam Alerts." *Cybercrime Support Network*, Guidestar Silver Transparency, June 21, 2021, *https://cybercrimesupport.org/covid-19-scam-alerts/*.

groups.[3] Although a wealth of research exists assessing the persuasive emotional methods used in phishing attacks, the tendency of users to fall victim to a variety of social engineering tactics might operate alongside the equally important factor of insufficient user awareness.

Given the increasing danger for both individuals and organizations, we must consider the various threats posed by online attacks across multiple spaces and from many different angles. Additionally, with technology an intrinsic part of our everyday lives, we need to be aware of digital threats, and be able to identify other threats such as phishing, ransomware, and spyware before they can wreak havoc in our professional or personal lives.

Throughout this book, we will discuss cyber threats such as phishing, disinformation versus misinformation in a post-COVID-19 world, cyber terrorism, and the power of social media, and ever-evolving privacy concerns in response to technological advances.

---

[3] Sanger, D. E., and Perlroth, N. "More Hacking Attacks Found as Officials Warn of 'Grave Risk' to U.S. Government." *The New York Times*, June 15, 2021, *https://www.nytimes.com/2020/12/17/us/politics/russia-cyber-hack-trump.html*.

CHAPTER ONE: THE SPIKE IN PHISHING AMID THE COVID-19 PANDEMIC

Key terms for this chapter

- **Authentication:** Verifying the identity of a user or process
- **Authority**: A sense of having or being in a position of control
- **Chaos theory**: An interdisciplinary theory stating that within the apparent randomness of chaotic complex systems, there are underlying patterns, interconnectedness, constant feedback loops, repetition, self-similarity, fractals, and self-organization
- **Initial condition**: A value of an evolving variable at some point in time designated as the initial time
- **Mail headers**: In an email, the body (content text) is always preceded by header lines that identify particular routing information of the message, including the sender, recipient, date, and subject
- **PHI**: Private health information
- **Phishing**: The fraudulent attempt to obtain sensitive information or data, such as usernames, passwords, and credit card details, by disguising oneself as a trustworthy entity in an electronic communication. Examples include:

1: The spike in phishing amid the COVID-19 pandemic

- o *Email phishing* – Using email to extract personal information from a target
- o *Spear phishing* – Attempting to extract personal information by targeting specific individuals
- o *Whaling* – Attempting to extract personal information by targeting specific high-end individuals
- o *Smishing* – Phishing via SMS/text message
- o *Vishing* – Phishing via telephone
- o *Angler phishing* – Phishing using social media

- **Urgency**: Importance requiring swift action
- **Verification**: The process of establishing the truth, accuracy, or validity of something

Phishing is of great concern in the public and private sectors all over the world. Recent examples include the COVID-19-related cyber attack that struck the United States Department of Health and Human Services; the email-detonated ransomware that prevented medical staff from accessing patient records,[4] which in turn led to an increase in heart attack fatalities; and the 2020 Iranian CARROTBALL malware campaign that spread via spear

---

[4] Akpan, N. "Ransomware and Data Breaches Linked to Uptick in Fatal Heart Attacks." *PBS*, October 24, 2019, *https://www.pbs.org/newshour/science/ransomware-and-other-data-breaches-linked-to-uptick-in-fatal-heart-attacks*.

phishing.[5] Both the public and private sectors experienced a doubling of attempted phishing attacks in 2020. There needs to be an increasing emphasis on user security awareness training, examining the technical means of users in at-risk industries.[6] As health care continues to represent the largest at-risk sector for cyber attacks in the US, government health agencies face a significant threat as holders of private health information. Malicious, foreign actors, as well as domestic attackers, desire access to this information to sell on the Internet's underbelly, known as the dark web.[7] Whereas previous research on phishing has focused on attacker persuasion methods, such as fear and urgency, and user responses to said methods, this book will focus on expanding users' knowledge of and tendency to verify email sender legitimacy in order to prevent successful phishing attacks.

This chapter will utilize chaos theory's argument involving the order existent in apparent randomness to analyze how insufficient user technical awareness could contribute to successful phishing attacks. Although other research has assessed chaos theory applications to various phishing

---

[5] For more information, visit:
https://www.forbes.com/sites/zakdoffman/2020/11/12/forget-russia-iranian-hackers-behind-malicuous-new-cyber-attacks-warns-new-report/?sh=1b200f3309ae.

[6] *https://apwg.org/trendsreports/*.

[7] Allodi, L., et al. "The Need for New Antiphishing Measures Against Spear-Phishing Attacks." *IEEE Security & Privacy*, vol. 18, no. 2, September 20, 2019, pp. 23–34, doi:10.1109/msec.2019.2940952.

attack methods, this book adopts this general theory to distinguish whether attacker methods or user knowledge and wherewithal account for the majority of successful phishing attempts. Thus, this book expands upon existing phishing research by investigating the following two factors:

1. User awareness regarding email header verification techniques
2. User tendency to verify email headers for signs of spoof[8]

If user knowledge of email header verification techniques leads to a decreased risk of engaging with a malicious email, this book's hypothesis holds, that users who are more knowledgeable about header verification will be less susceptible to emotionally persuasive phishing methods.

Chapter 2 explores the impact of federal health care's employee awareness regarding email sender verification looking at the following areas:

a) The risk of phishing to federal health care agencies
b) Machine learning tools for phishing prevention
c) Persuasive factors used in phishing attacks
d) User awareness of email legitimacy verification techniques

---

[8] For more information, visit: "What is Email Spoofing?" *Proofpoint, https://www.proofpoint.com/us/threat-reference/email-spoofing.*

e) Applications of chaos theory in cyber threat mitigation, including phishing attacks

Chaos theory in cyber threat mitigation

Although renowned for its association with the infamous butterfly effect, [9] as well as various applications within mathematics and theoretical physics, chaos theory initially emerged as a distinct branch of discrete mathematics that explores the multiple outcomes and effects a system can experience and produce as a sum of different inputs. First posited by French polymath Henri Poincaré, chaos theory holds that, although a system may appear chaotic upon first glance, its functions and products actually result from a multitude of factors that can continue to predict the system's output for a variable length of time.

In summary, a multitude of disciplines involve systems that, despite apparent randomness, adhere to an initial set of conditions. In addition to mathematics and astrophysics, chaos theory has been applied in psychology, economics, and even to the weather. Following the discovery of chaos theory, American mathematician Edward Lorenz postulated the concept of attractors, or a set of numerical values toward which a system evolves for a wide variety of starting conditions.

For instance, nonlinear phenomena, such as those present in dynamical systems across physics and engineering,

---

[9] For more information, visit:
https://www.americanscientist.org/article/understanding-the-butterfly-effect.

typically find expression in quadratic equations with foundations originally established by Lorenz. Because of the dimensions of three or more, chaotic systems become increasingly more challenging to detect patterns. For this reason, scientists from multiple disciplines have sought to identify a method to the madness. Ultimately, the chaos begins to clear once a pattern is determined. Historically, such patterns – called bifurcations – have emerged in the form of the swing of a pendulum and the brushless DC motor system, among other systems.

More recently, chaos theory has found relevance in various sectors of computer science and information technology, such as cybersecurity. Given the unpredictability of phishing techniques and user response, the application of cybersecurity could benefit security researchers in identifying patterns to make sense of which persuasion methods work best.

Another application of chaos theory within information security involves detecting patterns in authentication and cryptographic processes. Because of the inherent similarity between chaotic systems and the randomness of cryptographic keys used to secure private communications online, chaos theory has been used to assess the effectiveness of phishing attack detection. These methods are utilized to analyze both the specifics of these attacks as well as the systems already in place to help prevent phishing. In that sense, we can tackle the challenge of phishing by considering patterns used by attackers as well as the effectiveness of security controls, such as data loss prevention applications for email, such as Proofpoint.[10]

This investigation begins by examining the process of a generic phishing attack, including:

1) A fraudulent website that resembles a real website
2) Attacker sends a link to the fraudulent website
3) Victim inputs personal information on a malicious landing page, such as a fake Office365 login page credential harvester
4) Attacker abducts victim credentials

The monitoring process noted above can help researchers study how a phishing attack takes place, particularly regarding how target users respond at certain stages of the attack.

The application of chaos theory to the assessment of phishing attacks

In the area of phishing, the "Tools for Investigating the Phishing Attacks Dynamics" study used the website

---

[10] For more information, visit:
https://www.proofpoint.com/us/company/about.

PhishTank to pull data for analyzing the number of recorded phishing websites and historical attacks.[11]

Additionally, the number of attacks per given period of time were assessed, in particular the quantity of daily attacks within one month.

The study worked on the following assumptions:

> *"a) A system is steady if the observed deviations from a linear trajectory remain small*
>
> *b) A system is unstable when a sharp change in behavior from the baseline trajectory occurs"*[12]

In this specific study, the rate of verified attacks proved highly similar to suspected attacks, thus indicating the effectiveness of phishing attack prediction.

Although the previous applications demonstrate the usefulness of chaotic systems across various facets of cybersecurity and even phishing detection specifically, existing research has yet to explore the role of user awareness versus the effectiveness of phishing attack

---

[11] Lyashenko, V., et al. "Tools for Investigating the Phishing Attacks Dynamics." *2018 International Scientific-Practical Conference Problems of Infocommunications. Science and Technology (PIC S&T)*, October 9, 2018, doi:10.1109/infocommst.2018.8632100.

[12] Lyashenko, V., et al. "Tools for Investigating the Phishing Attacks Dynamics." *2018 International Scientific-Practical Conference Problems of Infocommunications. Science and Technology (PIC S&T)*, October 9, 2018, doi:10.1109/infocommst.2018.8632100.

methods through the lens of chaos theory. Therefore, we will explore this approach to investigate the role of persuasive attack methods versus levels of user technical knowledge of email sender verification tactics in terms of the impact of successful phishing attacks.

In addition to the general reference to chaos theory for establishing user security awareness as an initial condition that can help predict user behavior regarding phishing emails, socio-technical systems theory also comes into play via the necessary interaction between humans and computers. Socio-technical systems theory maintains that a complex organizational structure works best when society and technology collaborate as an interdependent whole. This theory remains highly relevant to cybersecurity, wherein humans must submit to routine training – such as for phishing preparedness – in order to efficiently understand how to interact with their systems and adequately protect data.

Because of the fact that sensitive data most frequently constitutes company assets and personal employee information, the secure management of that data often requires successful interactions between humans and machines. Moving beyond simply learning how to use a computer, users must also be trained in how to handle the data of both their fellow employees as well as the organization at large.

In order to supply teams with the ability to safely interact with computers as well as each other, managers can benefit by prioritizing user security awareness training. Useful training will educate employees on how to recognize cyber attacks such as phishing, namely by elaborating upon the traditional lessons in scoping for misspellings to include aptitude in analyzing email headers for evidence of sender address spoofing.

As email authentication headers show the recipient who sent the email and are available as a free add-on in the enterprise deployment of Microsoft Office 365, an assessment of the sender verification tactics and specifically spoof detection knowledge will serve as an initial condition for the potentially chaotic system of phishing attacks. Seeing as the action of header verification, even when a user possesses the necessary cognizance to interpret the headers in question, requires extra time and effort, we will consider email header authentication adherence as an additional initial condition within the dynamical system of phishing attacks. True to the nature of initial conditions in a dynamical system, email sender verification or lack thereof could present a significant determinant in the success of a phishing attack, even regardless of the various persuasive methods used. In particular, persuasive tactics and email header authentication serve as the primary modes of bifurcation throughout the chaotic system of phishing attacks.

Thus, this book posits user awareness surrounding email header verification techniques as a primary variable in assessing user susceptibility to phishing attacks across the federal health sector.

CHAPTER TWO: A GLANCE AT THE HISTORY OF PHISHING MITIGATION PRACTICES

A significant amount of research exists on the persuasive methods used to phish targets, as well as various technological tools developed to detect and prevent such attacks. For the purpose of this book, the following sections categorize the existing studies using the following themes:

- The risk of phishing to federal health care agencies
- Machine learning tools for phishing prevention
- Persuasive factors used in phishing attacks
- User awareness of email legitimacy verification techniques

The risk of phishing to federal health care agencies

Research at the Columbia University School of International and Public Affairs examined cyber attack activity following alleged Russian meddling in the 2016 French and US presidential elections. In particular, the research explored a spear phishing campaign conducted via social media that targeted US Department of Defense employees. Click rates of hyperlinks received during this campaign averaged 70%.[13]

---

[13] Dukarm, C.; Dill, R.; Reith, M. (2019). *Proceedings of the European Conference on Cyber Warfare & Security*. ACPIL, p172–7.

Although such a study from an Ivy League university presents useful insight into spear phishing against US federal employees, researchers can delve further into the reasoning behind the success of such campaigns by assessing the wherewithal of victims in terms of knowing when to refrain from engaging with artifacts such as an attachment or link embedded in an unsolicited email. Ultimately, this book provides a glance at the focus thus far on spear phishing attempts against US federal health care agencies specifically.

This chapter explores the US Department of Defense's trials in improving user security awareness and ability in the mitigation of risk compromise via phishing emails. As a large organization, the Department of Defense has sought to streamline administrative controls such as user training, and technical controls such as automated filters and warning messages, into an effective protection framework. Ultimately, this framework involves focused training objectives, a Department of Defense-specific content-sharing platform and a realistic delivery method for said content sharing.

Although the United Government and European Conference on Cyber Warfare and Security constitute reliable sources of research in cybersecurity, the user awareness training utilized in this initiative does not appear to delve into the users' tendency to verify a sender. Similarly, the training does not seem to prioritize educating users to adopt such verification practices. Therefore, an area of further research could investigate how users respond to a training method that focuses on instructing users to confirm sender legitimacy before any engagement with a suspicious email.

Anomali Threat Research identified a pervasive credential theft campaign targeting international and US federal agencies, as well as government procurement services.[14] Targeted organizations included the US departments of Commerce, Energy, Transportation, and Veterans Affairs, among others. The primary attack method relied on phishing emails with links that redirected to false login portals, hosted on website domains located in Romania and Turkey.

Although Anomali represents a reliable cybersecurity solution, which includes research capabilities, this study emphasized attacker methods rather than victim at-risk levels for susceptibility. As such, further research could build upon this study by assessing how employees of the victim agencies responded to this attack campaign. Overall, however, this research provides helpful insight into a previous phishing attack against US federal agencies.

An advanced persistent threat (APT) involves penetration into an organization's infrastructure, wherein the attacker establishes a backdoor or some other means to monitor the internal data and daily activity of the target system. As APTs account for a large number of nation state actor-sponsored cyber attacks against both the public and private

---

[14] Franscella, J. Anomali. "Threat Research Team Identifies Widespread Credential Theft Campaign Aimed at U.S. and International Government Agency Procurement Services." (December 12, 2019). *Marketwired*, p. NA.

sectors in the US, an assessment of the various threat vectors that enable access becomes pertinent.[15]

As one such intrusion method for APT involves spear phishing, or the targeted social engineering of certain personnel within an organization, federal personnel of high importance within federal agencies represent a prime risk. Indeed, with the increasing prevalence of bring your own device (BYOD) policies for mobile management in the workplace, the susceptibility of government personnel to vishing (phish via voice messages or phone calls) and smishing (phish via SMS) should constitute significant points of analysis for vulnerability programs within the public sector.

Although recommendations for mobile device management policy stringency in the context of information security at large do exist, a focus on training employees to examine email headers via mobile user interface has yet to be explored in any depth. In particular, prioritizing security awareness for government health care personnel specifically remains to be emphasized.

As phishing attacks provide the opportunity to evade antivirus software as well as intrusion prevention and detection systems by bundling malicious artifacts into

---

[15] Bann Lap, L., et al. "Trusted Security Policies for Tackling Advanced Persistent Threat via Spear Phishing in BYOD Environment." *Procedia Computer Science*, vol. 72, 2015, pp. 129–36, doi:10.1016/j.procs.2015.12.113.

seemingly harmless emails, phishing continues to present a significantly insidious threat to data security.

In fact, research reveals that antivirus detectors show greater accuracy when pinpointing malicious files than URLs.

Furthermore, an attacker can easily escape the executability of a link by bracketing out the domain segment and instructing the user to input the URL into a browser in order to better ensure execution.

Although this attention to the drawbacks of antivirus software denotes how such detection software might miss the mark in terms of phishing, existing research provides less visibility into antivirus success rates within the public sector specifically.

Similar to the general concept behind chaos theory of determining a method to the madness, various research exists portraying cyber attack methodology as an intricate web of various strategies. From among these strategies, nation state sponsored and other types of foreign threat actors use phishing as a primary tactic in evading intrusion detection and antivirus capabilities of organizations. By utilizing the human manipulation factor to conduct social engineering, attackers circumvent technological security controls by persuading authorized users to detonate malicious executables in the target environment.[16]

The magnitude of the cyber attacks that led to successful foreign interference in the 2016 US presidential elections, highlights federal susceptibility to disinformation and social engineering. Despite this vulnerability to user-level manipulation, in-depth analysis and reform of existing technical user security awareness training in the federal space remains to be investigated.

In particular, an even less explored area of social engineering is the category of spear phishing known as whaling – or phishing targeting high-level personnel. Given the influence and degree of privileged access possessed by so many agency directors, especially in the federal space, attackers stand to gain substantial benefit by manipulating these types of users. Indeed, threat actors could potentially utilize whaling for system and network intrusion across many areas of the federal sector – from weapons sales to technology trade, and development to aerospace.

Similar to the topic of the role of phishing in the 2016 elections interference, the issue of whaling in the federal sector and particularly ensuring adequate user training

---

[16] D., Gioe V. "Cyber Operations and Useful Fools: The Approach of Russian Hybrid Intelligence." *Intelligence and National Security*, vol. 33, no. 7, December 28, 2018, pp. 954–73, doi:10.1080/02684527.2018.1479345.

toward mitigation therein could benefit from further research.

Applications of chaos theory in cyber threat mitigation, including phishing attacks

The "Tools for Investigating the Phishing Attacks Dynamics" study may represent one of the initial studies conducted into the application of chaos theory to cybersecurity and phishing specifically. The aforementioned study analyzes trends in phishing attack methods drawn from various threat hunting engines in order to try and find a pattern. Whereas such research uses the mathematical aspect of chaos theory to make predictions on future phishing attack methods, this book utilizes the overarching concept of initial conditions in chaotic systems to explore how user technical knowledge might account for a little-studied initial condition that factors into the handling of phishing attacks.[17]

Montero-Canela's research represents a study of chaos-based cryptosystems for ad hoc networks, perhaps the most common application of chaos theory to the cybersecurity

---

[17] Lyashenko, V., et al. "Tools for Investigating the Phishing Attacks Dynamics." 2018 International Scientific-Practical Conference Problems of Infocommunications. Science and Technology (PIC S&T), February 4, 2019, doi:10.1109/infocommst.2018.8632100.

arena.[18] Although chaos theory provides a highly useful framework for developing the field of cryptography, this book focuses less on mathematical applications and more on the initial condition aspect of user awareness in the face of phishing threats.

In the case of a phishing attack, the various factors attributed to levels of user susceptibility have traditionally centered on emotional tactics employed by attackers, as well as the personality traits of the victims. However, this book will shift away from emotional aspects to focus on potential gaps in user technical knowledge. Against the backdrop of research conducted on the results of emotion-based attacker methods, user email authentication awareness will emerge as another potential initial condition to phishing risk.

Although human-to-computer interaction remains essential for systems engineering [19] and various design methodologies highlighted in Baxter and Sommerville's, 2011 "Socio-technical systems: From design methods to systems engineering, interacting with computers" study, we will focus specifically on the aspect of the human-computer interface involving user awareness and aptitude

---

[18] Montero-Canela, R., et al. "Fractional Chaos Based-Cryptosystem for Generating Encryption Keys in Ad Hoc Networks." *Ad Hoc Networks*, vol. 97, February 2020, p. 102005. Doi:10.1016j.adhoc.2019.102005.

[19] Baxter, G., & Sommerville, I. (2011). "Socio-technical systems: From design methods to systems engineering, interacting with computers," 23(1), 4–17.

at analyzing email headers. Similar to the 2011 study, this book utilizes user awareness surrounding email verification as a primary indicator of successful mitigation against attempted phishing attacks. Similarly, Baxter and Sommerville's research examines the practical application of socio-technical systems theory to cybersecurity as a whole. Indeed, the human-to-machine focus of this theory effectively caters to the value of user security awareness for the safe utilization of computers.[20] True to the treatment of email verification knowledge as a determinant for phishing mitigation, socio-technical theory provides a strong framework with which to combine the view of the phishing attack process as a chaotic system.[21]

Finally, we'll explore the Microsoft email header verification framework in order to assess the technical knowledge of a group of users from the target federal demographic. Analyzing the ability of users to distinguish between legitimate and spoofed or otherwise malicious emails based on message headers, represents a socio-technical lens of phishing research that places the power and confidence back in the hands of the human user.

---

[20] Malatji, M., et al. "Validation of a Socio-Technical Management Process for Optimising Cybersecurity Practices." *Computers & Security*, vol. 95, 2020, p. 101846, doi:10.1016/j.cose.2020.101846.

[21] Zimmermann, V. "Moving from a 'Human-as-Problem' to a 'Human as-Solution' Cybersecurity Mindset." *International Journal of Human-Computer Studies*, vol. 131, 2019, pp. 169–87, doi:10.1016/j.ijhcs.2019.05.005.

Therefore, by considering both an organization's attention to training users how to interpret these email headers, as well as the current extent of these users' depth of knowledge surrounding these headers, this book seeks to comment on the present phishing preparedness among employees in the federal health care sector.

Machine learning tools for phishing prevention

Because users at the majority of organizations receive a plethora of daily emails, machine learning approaches to the prevention and detection of spam emails and phishing attacks have been undertaken. Literature on this topic echoes the aim of this book in its emphasis on email header analysis. [22] However, whereas, for example, Kadam and Morovati focus on an entirely automated algorithm for determining malicious emails, this book will examine the potential for a hybrid human-as-solution and machine-assisted phishing prevention strategy. Their "Detection of Phishing Emails with Email Forensic Analysis and Machine Learning Techniques" study stands out in its attention to emails that bypass spam filters. This book adopts a similar skepticism of automated filters by suggesting the implementation of an application that notifies users of potential signs of phish in real time, allowing the user to verify the flagged element's degree of maliciousness.

---

[22] Morovati, K., & Kadam, S. S. (2019). "Detection of Phishing Emails with Email Forensic Analysis and Machine Learning Techniques." *International Journal of Cyber-Security and Digital Forensics*, 8 (2), 98+.

A new framework approach to cybersecurity and specifically phishing combines the artificial intelligence and machine learning CANTINA (DMLCA) with the software defined network paradigm to streamline phishing defense capabilities.[23] These combined programs help to automate and improve the process of accurately identifying the threat level presented by both hyperlinks and documents contained within suspicious emails. Similar to Kadam and Morovati's research, Raja and Ravi's study focuses more on the machine learning aspect than on how artificial intelligence can be combined with human security awareness to help mitigate phishing attacks.

Although a substantial development in cybersecurity innovation toward phishing detection, this book could be expanded by examining use cases across various industries. That is to say, this software could be tested to detect phishing attempts in real-time in both the public and private sectors, in order to provide more insight into the applied success rate of this novel technology. On the whole, this study promises an increasingly innovative application for machine learning in the cybersecurity arena.

Persuasive factors used in phishing attacks

In the face of the nigh unprecedented atmosphere of the COVID-19 global pandemic, security teams across the

---

[23] S., E. Raja, and Ravi R. "A Performance Analysis of Software Defined Network Based Prevention on Phishing Attack in Cyberspace Using a Deep Machine Learning with CANTINA Approach (DMLCA)." *Computer Communications*, vol. 153, March 1, 2020, pp. 375–81, doi:10.1016/j.comcom.2019.11.047.

public and private sectors alike have scrambled to prepare their employees for the increasingly ubiquitous number of urgency-driven scams playing on the nerves of anxiety-ridden users. In fact, various scams have specifically targeted US federal agencies, such as the Department of Health and Human Services. Thus far, many of the attacks exploiting fear over the COVID-19 situation have relied on phishing tactics, including persuading users to engage with malicious URLs purported to lead to health care websites or landing pages with news on the pandemic.[24]

Despite the prevalence of these COVID-related attacks, research into user technical awareness of email verification tactics remains scarce.

Indeed, the frightened atmosphere wrought by COVID-19, as well as remote work protocol situations wherein employees sometimes lack easy access to the advice of security personnel, have resulted in a plethora of successful phishing attacks. These attacks have granted many threat actors the credentials of users from numerous organizations across the United States, including entities in the federal space.

Presently, the majority of literature on the user impact of time-sensitivity and urgency in emails examines how such persuasive tactics frequently aim for user engagement via the presentation of something to gain – namely, money, or, in the case of COVID-19, allegedly crucial health

---

[24] Chapman, P. (2020), "Are your IT staff ready for the pandemic-driven insider threat?" *Network Security* (1353–4858), 2020 (4), 8.

information.[25] Even so, this book directly references the scarcity of existing research on how urgency can influence user susceptibility to phishing attacks.

Even before the onslaught of urgency and authority-based phishing attacks during COVID-19, attackers have frequently adopted an urgent tone in their emails. As most humans naturally react to a sense of urgency with a degree of adrenaline generated by the amygdala – the area of the brain that regulates emotional response – an expected reaction to a time shortage involves rapid response to whichever action depends on the deadline in question.[26] For instance, in the case of a URL leading to a website on pandemic facts that is about to expire, a user might wind up clicking on the link before taking the time to verify the email sender. The "Time Pressure in Human Cybersecurity Behavior: Theoretical Framework and Countermeasures" study explores the correlation between social engineering tactics and target personality. More specifically, the investigation examines the factors of email phishing that might predict user susceptibility. Overall, the research

---

[25] Marks, J. (2020). "The Cybersecurity 202: Coronavirus pandemic unleashes unprecedented number of online scams." *The Washington Post*, *https://www.washingtonpost.com/news/powerpost/paloma/the-cybersecurity-202/2020/04/01/the-cybersecurity-202-coronavirus-pandemic-unleashes-unprecedented-number-of-online-scams/5e83799b88e0fa101a757098/*.

[26] N., Chowdhury H., et al. "Time Pressure in Human Cybersecurity Behavior: Theoretical Framework and Countermeasures." *Computers & Security*, vol. 97, October 2020, p. 101963, doi:10.1016/j.cose.2020.101963.

detected a connection between high extroversion and a higher degree of risk regarding phishing emails. It also found that the users in question seemed on high alert, exhibiting greater accuracy when identifying a phishing email as phishing than when labeling a legitimate email as legitimate. In fact, more than 54% of participants inaccurately labeled emails as phishing.

Research has shown that when examining phishing activity and its subsequent success, specialists should bear in mind both the emotional appeal tactics used – such as authority, likability, and urgency – as well as the emotional state of the targeted individual at the time of the attempted attack. A study in the *Personality and Individual Differences* journal notes the lack of current research into the emotional state of the user in question, citing the importance of factors such as distraction, fatigue, or haste in an individual's decision to interact with a malicious email.[27]

Although this study perhaps comes closest to the planned book topic of user wherewithal regarding sender verification at US federal agencies, the research present here does not focus on the public sector nor on the likelihood of sender verification. Therefore, this book provides a lucrative catalyst for studies assessing the tendency of federal agency employees to verify senders for fraud and spoofing.

---

[27] G. Norris, and A. Brookes. "Personality, Emotion and Individual Differences in Response to Online Fraud." *Personality and Individual Differences*, January 8, 2020, doi:10.1016/j.paid.2020.109847.

2: A glance at the history of phishing mitigation practices

The *Cyberpsychology, Behavior, and Social Networking* journal published a study that specifically examines phishing susceptibility in use cases involving the persuasive tactics of authority and likability.[28] Contrary to expectations, susceptibility to phishing and fraud actually decreased as the sense of urgency increased. Additionally, phishing susceptibility rose in response to higher levels of perceived authority and likability.

As this book will investigate user tendencies to verify email sender for spoofing activity when faced with a variety of emotionally persuasive tactics, this research provides an optimal example of how to carry out such an assessment. However, this study lacks a focus on US federal agency high-level personnel, as well as user wherewithal regarding sender verification.

User awareness of email legitimacy verification techniques

With many services, such as education, moving online during the COVID-19 global pandemic, the FBI advised all computer users to remain vigilant regarding an increase in cyber attacks, specifically phishing campaigns.[29] With the prevalence of attackers adopting methods such as urgency to persuade young and older users alike into falling for

---

[28] Baryshevtsev, M. (05/2020). "Persuasive Appeals Predict Credibility Judgments of Phishing Messages."
Cyberpsychology, Behavior, and Social Networking (2152–2715), 23 (5), 297.
[29] Haro, F. (2020, May 12). "FBI warns against COVID-19 phishing scams." *UWIRE Text*, p. 1.

scams and otherwise granting access to their computer systems, the US government has warned that a distressing experience such as COVID-19 warrants even greater attention to be given to who exactly the sender of emails is.

This emphasis on the attacker's use of urgency in cyber attacks, such as phishing to play on user emotions at a time of heightened stress, plays well into the need for further investigation into a current consensus on the ability of individuals to think critically and question the email sender in such situations. Moreover, a federal agency such as the FBI prioritizing emotionally persuasive techniques employed during phishing attempts elucidates the importance of studying these tactics in more depth.

In terms of research on user security awareness training, various studies exist worldwide. One such study from Thailand focuses on a preference for classroom versus online training, rather than which method or type of content seems most impactful to phishing prevention.[30] In fact, this particular experiment has revealed that the mode of security training yielded an overall insignificant decrease in successful phishing attacks.

In light of increasingly sophisticated spear-phishing tactics and an overall lack of sufficient user awareness training throughout the globe, the need for innovative phishing mitigation measures has come to light. Indeed, "The Need

---

[30] K., Tschakert Florian, and Ngamsuriyaroj S. "Effectiveness of and User Preferences for Security Awareness Training Methodologies." *Heliyon*, vol. 5, no. 6, June 2019, doi:10.1016/j.heliyon.2019.e02010.

for New Antiphishing Measures Against Spear Phishing Attacks" study reports that more than 76% of organizations from a sample spanning 32 countries have fallen victim to spear-phishing attacks.[31] In general, because of the more personalized nature of spear phishing, the attacker typically performs thorough research tailored to the victim rather than an adding an overarching sense of urgency or scarcity. For instance, the attacker might craft an email mentioning a relative or specific job position of the user in question. This level of nuance makes a case for closer attention to detail when assessing an email body for legitimacy, particularly serving as an example of a scenario in which automated phishing detection may not suffice.

In terms of attack assessment rather than user preparedness, the phishing prevention and awareness training service, Cofense, has established an up-to-date database of the latest phishing attacks to bypass email security controls.[32] Similar to automation and machine learning tools developed to prevent and detect phishing, Cofense's database serves to provide organizations of various industries with a reference base for the various URLs,

---

[31] L., Allodi, et al. "The Need for New Antiphishing Measures Against Spear Phishing Attacks." *IEEE Security & Privacy*, vol. 18, no. 2, September 20, 2019, pp. 23–34, doi:10.1109/msec.2019.2940952.

[32] Cofense Launches Free Resource Center and Searchable Database Highlighting the Latest Phishing Attacks that Bypass Email Security Technologies. (June 9, 2020). *UWIRE Text*, p. 1.

fraudulent website names, and filenames used in previous attacks.

Unfortunately, the 2020 compromise of consumer credit company Experian, proves that increased investment in heightened information security measures does not always guarantee a decrease in successful attacks. In fact, this particular case has revealed that only 23% of respondents surveyed stated confidence in their ability to recover from damage to reputation and trust relationships with business partners. In fact, 66% of respondents admitted to not having reviewed or updated their organization's information security policy since the time of the policy's instatement. [33] Thus, the general complacency of organizations across many industries suggests that many entities could benefit from revisiting the priority of data security.

Despite the general scarcity of user security awareness training that focuses on technical steps toward identification of email sender spoof, the "Countermeasure Technique for Email Spoofing" study emphasizes a specific process for email authentication. This process entails verification of email headers, with particular focus

---

[33] New Osterman Survey, "Phishing Prevention Perception Gap Reveals Disconnect Between C-Suite and Cybersecurity Professionals," (March 2, 2020). *PR Newswire*, p. NA.

on a comparison between the "From" field and "return path" or "sender-envelope."[34]

In terms of existing literature detailing the technical steps for email verification, this study comes closest to the framework used in this book. Moreover, the fact that this framework has been used in a previous assessment of user awareness in email security, indicates that the element of sender authentication serves as a lucrative standpoint for the potential ability of technical knowledge to help mitigate against phishing attacks.

This book utilizes the concepts of initial conditions in chaos theory to conduct a largely qualitative study of the role user technical awareness plays in the dynamical system of phishing attacks and mitigation. As such, many studies discuss useful examples of how to identify patterns in a seemingly chaotic system. For instance, the swinging of a pendulum or changes in the weather over the period of a day.[35]

Overall, the underlying goal of viewing a process through the prism of chaos theory remains to identify those starting factors that can end up producing significant impact upon the process as the domino effect continues onward. In the

---

[34] Kumar, M., Hanumanthappa, M., Kumar, S., T. V.. A "Countermeasure Technique for Email Spoofing. International Journal of Advanced Research in Computer Science," Jan/Feb 2013, Vol. 4, Issue 1, p128–33, 6p.

[35] Yang, F., et al. "Sequence of Routes to Chaos in a Lorenz-Type System." *Discrete Dynamics in Nature and Society*, vol. 2020, 2020, pp. 1–10, doi:10.1155/2020/3162170.

case of phishing attacks and mitigation, the initial condition of user awareness regarding email verification techniques could stand to present such a lasting effect.

Chaos theory has also found application in the realm of information security via the complex development of cryptographic systems. My own independent study focuses on the fact that, despite an inherent desire for randomness in the devisement of crypto keys, the concept of chaos theory has yet to be practically applied to the field of cryptography in a manner that has proven useful to securing a working key exchange system.[36] Presently, the majority of research in this area centers on using chaotic maps to determine sensitivity to parameters and loosely plan out the network topology of cryptographic systems used by the Internet of Things (IoT) networks.

Another overarching trend across the various existing sources on the use of both email verification knowledge and chaos theory in the study of phishing attacks, concerns country of origin – namely, the majority of studies on the application of both user training in email authentication and framing phishing attacks as a chaotic system stem from research conducted across a variety of countries. Therefore, this book stands to add to the existing literature in terms of contributing an analysis of user email security awareness among US federal personnel.

---

[36] J., I Sen, et al. "Implementation and Practical Problems of Chaos-Based Cryptography Revisited." *Journal of Information Security and Applications*, vol. 50, February 2020, p. 102421, doi:10.1016/j.jisa.2019.102421.

CHAPTER THREE: THE HARD TRUTH – HOW LITTLE MANY USERS UNDERSTAND ABOUT IDENTIFYING MALICIOUS EMAILS

After spending most of my professional career in incident response targeting phishing attacks, I realized an unfortunate pattern of non-technical coworkers across multiple industries – from finance to telecommunications to government – didn't really understand much of what they were looking at in terms of email headers. In fact, many didn't even feel a need to distinguish between spam and phishing, resulting in a daily flurry of spam to the security team's review queue.

This knowledge gap inspired me to take a chance and actually try and see just how much these individuals understood regarding the slew of emails received in their inbox.

Along lines loosely similar to pattern identification in chaos theory, I crafted a survey to explore whether attackers were simply brilliant, employees didn't absorb their annual security training, or perhaps something else entirely – namely, said security training not going deep enough to detect spoofed emails that appear legitimate upon first glance.

Given that the health care sector represents one of the largest targets of cyber attacks since the start of COVID-19, I developed a survey to ask the following questions:

Do federal health care agency personnel have the knowledge and wherewithal to verify email authentication headers when choosing whether to engage with unsolicited emails?

This question builds into the following three questions that were posed to 50 respondents in the federal sector, with an emphasis in health care where possible:

1. **Upon receiving an unexpected email, are you aware of the ability to check the email headers to verify sender legitimacy?**

   - Yes
   - No

2. **If you are aware of the email header functionality, do you tend to try and check the email headers for every unsolicited email received?**

   - Yes
   - No
   - I know to check email headers, but my organization's mail client does not provide this option.

3. **Would you find it helpful to be notified in real-time whenever an email you are viewing is potentially malicious based on a lack of header match?**

   - Yes – I would like to know right away so I can look into it.
   - No – That sounds too complicated or time-consuming.

In addition to assessing technical awareness surrounding email authentication, subjects were consulted via a survey in order to assess whether users at federal health agencies who knew to check mail authentication headers were more likely to identify an email as malicious, despite the presence of persuasive phishing methods.

Although the risk of a user falling victim to persuasive phishing tactics seems random, knowledge of email header authentication techniques might be a strong variable in altering the outcome of targets choosing to engage with an unsolicited email.

Sample population

Subjects for this book's research were at large drawn from employees working in the federal sector at the following agencies: the Center for Disease Control and Prevention, the United States Department of Defense, and the United States Department of Health and Human Services. Because the aforementioned agencies represent the largest federal health agencies in the United States, they have been selected as the sources for the subjects of this research.

Out of all the users polled from across the private and public spaces of the health care sector, more than half of the respondents reported not being aware of the ability to analyze email headers for sender authenticity. Additionally, 100% of those users who acknowledged a lack of awareness regarding headers, attested to the benefit of an email security tool to assist with header analysis. Although the tested sample has its limitations, this book aims to set a foundation for further investigation into the human role in email security.

Tackling email headers

A case study

In order to blend the human element of user security awareness with automated phishing prevention, let us consider a hypothetical email header analysis tool called HeadCount.

HeadCount is a proposed software offering the latest capabilities in employee network activity surveillance, data loss prevention, and web content filtering. With mechanisms in place to ensure the domains of email sender fields match authentication headers, such a tool can provide users with a helpful head start in swiftly analyzing email headers, with a deeper nuance than an entirely automated algorithm.

Although companies worldwide have successfully deployed other email security tools, such as Proofpoint, we envision the familiar question of employee privacy to pose the greatest challenge in implementing a technology such as HeadCount. As every organization should prioritize employee autonomy, privacy, and security with the adoption of a BYOD policy, user privacy concerns surrounding inbox monitoring would be a top priority.

Challenges to implementation

When integrating a new technology into their enterprise environment, leaders face four primary challenges:

1. Administrative
2. Technical
3. Legal

4. User-based

Administrative challenges

On an administrative level, executive management and business unit leaders must devise policies and procedures for how the technology will be implemented amid existing technology and utilized by users. Specifically, the user aspect concerns training procedures for employees in orienting them to the extent of the technology's capabilities. Additionally, detailed policies should describe all the benefits of the technology, so that users understand the reasoning behind its adoption. Finally, the policy should also include limitations to the technology, such as corners of the scope that will remain unimpacted and unchanged.

Technical challenges

From a technical standpoint, all software and similar technological innovations will be implemented into the existing network and system topology. Therefore, versions of all new software will be compared with existing software within our environment in order to ensure compatibility. Moreover, new vulnerability assessment teams as well as penetration testers will analyze all new software for flaws and weaknesses in need of patching. As soon as a flaw is discovered, a new supply of that software will be requested and undergo a repeat assessment. In terms of BYOD policies, all devices from outside of the organization that employees bring to campus will undergo rigorous security assessment to ensure they are safe to operate on our networks. After all, considering BYOD has taken effect even more prevalently in Europe and the Middle East than in the United States, we would do well to monitor usage in that region as efficiently as possible.

Legal challenges

Regarding legality, any monitoring activities that our organization adopts for the purpose of user protection and browsing surveillance must first be cleared with the legal teams of all of our global branches before going into effect. Adding on to the necessity for copyright, patent, and similar issues of product trade, we envision the legal team serving as the primary point of consultation for issues of technology commerce as well as employee privacy rights. Particularly concerning the use of HeadCount, before deployment we will first ensure the legal teams in each of our domestic and international offices confirm that surveillance does not infringe upon employee confidentiality. Before the installation of HeadCount, all surveillance – including application (email included), network, and system logs will continue to be monitored by the IT staff at each separate branch, using existing and approved tools such as Office 365, Proofpoint and Splunk.

User-based challenges

Building upon management of technical and legal matters, the users operating under such legal policies and utilizing these technologies represent our chief priority. In protecting user rights, we will conduct monthly surveys to assess both how employees feel regarding the new technologies implemented with which they do not engage, as well as their level of proficiency from training with those that they use. In order to promote streamlined learning for all employees, training sessions on each newly adopted technology will be held for all members of every team at the same time.

Next steps: Computer-assisted threat detection

Ultimately, the hypothetical security tool HeadCount, as well as this book, serve to further research into how users working in the industry most targeted by cyber attacks can collaborate with software to verify email legitimacy more efficiently and conveniently. The accompanying question of whether users would prefer software assistance in such verification, helps to develop a framework for further improvements in security infrastructure for the health care sectors across both the federal and private arenas.

All that said, HeadCount remains purely hypothetical, a mere symbol of a tool that management could use to help train users not only in recognizing the signs of a malicious email but also learning to work alongside an assistant software to avoid phishing attacks. The main takeaway here should be user awareness training combined with machine learning comprise a crucial catalyst for information security, and one element cannot function as capably alone as together.

By and large, with all of the false information circulating surrounding the COVID-19 pandemic, many have turned to social media both as a resource and scapegoat for gathering and spreading reliable updates on this global crisis. In the next chapter, we explore how criminal hacker activity and misinformation concerning political elections in various countries have plenty on edge in terms of whom to trust these days.

CHAPTER FOUR: DISINFORMATION VERSUS MISINFORMATION IN A POST-COVID WORLD

Because of the controversy surrounding foreign fake news that allegedly influenced the 2016 presidential elections in countries such as France and the United States, concern over the ubiquity of this misinformation has skyrocketed in the past two years. In particular, many social media platforms such as Facebook and Twitter, have received major backlash for failing to effectively protect their users by not monitoring fake news more closely, both in terms of health news as well as politics. Perhaps most interestingly, the need to differentiate between deliberate disinformation by mal-intentioned entities versus the more often accidental misinformation shared by misinformed individuals, has grown more apparent.

Global information sharing

Over the past two years, social media giants have placed particular emphasis on protecting users from fake news as well as graphic violence and pornography.

However, the well-known social media platforms, such as Facebook, Instagram, and Twitter, sport billions of users worldwide. Therefore, the practices employed toward fake news mitigation must account for the emergence of said news in a multitude of different languages. The trouble is, there are hardly enough content moderators to spot all of this unsavory material.

4: Disinformation versus misinformation in a post-COVID world

Unfortunately for the 2016 election era and potentially its 2020 counterpart, social media content moderation's prioritization of extreme violence and sex leaves ample opportunity for misinformation to slip through the cracks – especially when said information appears within seemingly legitimate news stories, often in languages that the majority of moderators do not understand.[37]

Content moderators

After verifying sources for any known blacklisted websites, a content moderator should take the next step of ensuring they can read the language in which the material appears. Even in regard to reviewing content in a language unknown to moderators, content moderation policy should implement the practice of checking for buzzwords in article titles using Google Translate.

Because of the clickbait nature of the following types of buzzwords:

- Names of politicians
- Killed
- Assassinated
- Hacked

---

[37] Earl, J. "Google's Top Search Result for 'Final Election Numbers' Leads to Fake News Site." *CBS News*, CBS Interactive, November 15, 2016, https://www.cbsnews.com/news/googles-top-search-result-for-final-election-numbers-leads-to-fake-news-site/.

- Election, etc.

content moderators should search the text in question for such terms in a variety of languages so that a foreign-language article will not pose such a moderation barrier. For languages written in different character systems, it is recommended that moderators take note of how they appear and use their notes as a reference point when perusing news story headlines in shared articles.

How to spot misinformation

As with determining the legitimacy of any web page, moderators are encouraged to check the news source websites with the following precautionary checklist in mind:

1. **Pop-ups:** Do multiple pop-ups and ad banners appear when trying to navigate the web page?
2. **HTTP vs. HTTPS:** In the address bar at the top of the web page, does the far left read **http** or **https**? **Tip:** Although not set in stone, https websites tend to be more secure as their data is encrypted.
3. **URL redirect:** Does the article link seem to redirect to multiple different URLs before the actual destination page loads? An optimal free tool for verifying the safety of URLs is VirusTotal.
4. **Hyperlink match:** Does the link in the address bar match the name of the website you are trying to reach?
5. **Typosquatting:** "Typosquatting is what we call it when people - often criminals - register a common misspelling of another organization's domain as their

own."[38] Examine the link for a typosquatted domain or URL. For example, consider **facbook.com** rather than the proper **facebook.com**.

Links become quite dangerous when they lead to a login portal that prompts you to enter your email or social media login credentials, after which point the criminal hacker will have access to your actual email or social media account. Once again, VirusTotal is an invaluable free resource to safely test suspect domains and URLs before opening them on one's machine.

Finally, when presented with a foreign-language news article while browsing either social media or the Internet in general, Google Translate must always be used with a pinch of salt.

Content moderators should *always* cross-reference any search terms with articles from reliable news sources in their native language, to ensure as best as possible that nothing is taken out of context. As certain idioms and slang can be easily lost in translation, one should never rely on Internet translation for more than one search term at a time.

Managing content moderation: In theory and practice

With the burgeoning dissemination of online information sharing and the use of social media platforms, three prominent social and ethical issues have arisen.

---

[38] *https://support.microsoft.com/en-us/topic/what-is-typosquatting-54a18872-8459-4d47-b3e3-d84d9a362eb0*.

4: Disinformation versus misinformation in a post-COVID world

1. How to effectively use technology to moderate online content
2. The emotional impact of graphic content on human moderators
3. Specific tools developed to moderate online material.

To this effect, multiple countries as well as governance bodies within organizations have ruled on ways to safeguard end users and organizations from the potential harm of illicit content.

Over the past four years and particularly since foreign meddling in the 2016 US presidential elections via social media, multiple automated and human-assisted artificial intelligence tools have arisen as businesses strive to protect their customers, employees, and reputations. [39] Additionally, various policies have emerged regarding how such content moderation tools should be implemented in a way that both defends users from online harm and upholds their right to freedom of expression.

In terms of theoretical approaches to leveraging technology toward content moderation, the question of who takes responsibility for such surveillance prevails. For instance, large social media companies such as Facebook and Twitter, tend to outsource their content moderators. Meanwhile, New York University, among other entities,

---

[39] Schaake, M., and Reich, R. "Election 2020: Content Moderation and Accountability." *Human-Centered Artificial Intelligence Cyber Policy Issue Brief,* Stanford University, October 2020.

has called out the social media giants on their practice of depending on contract staff for content moderation, a strenuous task that yields less pay and mediocre benefits compared with those of full-time employees. In addition to suggestions for moving content moderation in-house, these entities advocate for further research into the frequency of post-traumatic stress disorder (PTSD) that can occur among moderators, especially as this type of analysis requires a human nuance and can therefore unlikely be fully automated.[40]

However, as governing bodies in the technology industry and beyond aim to streamline the content moderation process, the goal of upholding free speech within a digital environment has been juxtaposed in the context of social media. As companies such as Facebook have grown to serve billions of users, content moderation of social media platforms has come to rely on computer algorithms to help determine what should be removed while also safeguarding users' rights to free speech. Notwithstanding, the definition of "freedom of expression" in the context of users' democratic rights often remains murky amid social media giants' reservations to disclose their algorithms and policies behind how they moderate content on their platforms.

In the face of such opaque definitions of content policy, governance bodies and legislation sometimes clash in

---

[40] Criddle, C. "Facebook Moderator: 'Every Day Was a Nightmare.'" *BBC News*, May 12, 2021, *https://www.bbc.co.uk/news/technology-57088382*.

discussions surrounding the efficiency of automated content moderation. In general, unsupervised artificial intelligence automation tools could render increasingly unclear the policies behind decisions on whether to leave or remove content from social media platforms.

To this end, Georgetown Law School has framed a set of guidelines for maintaining democracy in the arena of online content moderation. These guidelines center on establishing an administration and monitoring body to oversee content moderation practices, with the goal of holding users themselves accountable for their actions on a given online platform, and to have organizations reveal their moderation policies to the public.

Furthermore, the growing question of content moderation has placed increasing pressure specifically on US technology companies to crack down on their moderation policies. This need has been augmented by the Federal Trade Commission's investigation into Facebook for alleged antitrust activity, action that has been tied to general shortcomings in social media content moderation practices in terms of monitoring the nature of user interactions on the platform.

Aside from governance issues surrounding content moderation, the human moderators themselves feel the impact of daily perusal through often graphic online content. In fact, some moderators have even filed lawsuits following experiences of PTSD. Since 2016, several interviews have emerged with former moderators as well as company executives on how best to tend to the needs of these moderators when emotional trauma arises.

On the other hand, content moderators who sift through thousands of user-reported content per hour might even become desensitized to some of the material they see. This means that after the hundredth image of a beheading or thousandth video of sex trafficking, the moderators might actually start clicking through without bothering to report or deactivate the associated account.

In addition to issues surrounding the psychological impact of human moderators, the question of policy enforcement emerges. The *International Review of Law, Computers, & Technology* journal has issued a stance showcasing dilemmas with content moderation enforcement in practice. In particular, human rights concerns such as avoidance of law enforcement escalation and reporting factor into the scrutiny of many of these social media giants. Often, such disregard for the law enforcement's escalation of abuses, such as human trafficking, occurs to evade the perceived PR risk to a company's branding.

As many social media platforms outsource the task of content moderation to third-party vendors, various such contracting agencies have taken different stances on client policy disclosure. Cognizant is an example of a contracting company that has bowed out of assisting social media giants such as Facebook with content moderation. The leading cause behind the withdrawal involved content moderators experiencing psychological trauma over the reviewed material.

Moreover, on the topic of overall emotional toll on content moderators as individuals, research has shown that psychological exhaustion results in less accurate moderation. However, breaking up moderation tasks has proven to help alleviate this burnout rate.

Running parallel to the theoretical questions of content moderation enforcement planning, the viability of applied moderation technologies enters the conversation. Legislative bodies across Europe and the United States question the ability for analytical accuracy of automated content moderation tools. Namely, some US entities worry that such AI tools cannot be configured to perfectly adhere to policies that safeguard users' right to free speech under the First Amendment.

Next steps: Head above water in cyberspace

The good news is, thwarting these risks does not always require a technical pedigree – simply choose attention to detail and, most importantly, *understand* what one is looking *at* and *for*.

Cyberspace is a tricky terrain. New breeds of criminal hackers and threats are constantly evolving, and many flock to social media platforms to spread fake news and malicious links. The aforementioned tips are just a few pointers in a vast array of ever-changing techniques needed to ensure that both users and content moderators alike stay safe and properly informed. Perhaps the main aspect for which social media information security managers should remain vigilant are language barriers – namely, the sharing across their platforms of foreign-language news articles and similar content that could contain disinformation. As with the aforementioned phishing threat facing the health care industry, a starting point to tackle foreign language misinformation could begin with the implementation of software that alerts moderators to certain politically themed key words across multiple languages.

Indeed, alongside the ever-present threat of disinformation and misinformed users spreading inaccurate news, a potentially far more sinister use of social media also continues to rear its ugly head – cyber terrorism.

CHAPTER FIVE: CYBER TERRORISM AND THE FIREPOWER OF SOCIAL MEDIA

Upon hearing the term "cyber terrorism," one might think of China, Russia, or even the United States. Frequently, the ability to carry out threatening or harmful acts via the Internet is associated with countries considered to possess sophisticated technology – that is, technology capable of hacking into the security systems of foreign nations. However, such an association overlooks the power of a ubiquitous online tool – social media – namely YouTube videos, heavily utilized by radical groups such as the Afghan Taliban, as well as the more recent terror initiative Islamic State of Iraq and the Levant (ISIL), among other entities.[41]

Bruce Hoffman, Professor of Security Studies at Georgetown University, defines cyber terrorism as:

"The deliberate creation and exploitation of fear through violence or the threat of violence in the pursuit of political change."

Using this definition, "cyber terrorism" could constitute any threat of violence or war mediated via the Internet. In the particular case of YouTube and similar online video display platforms, Afghanistan – a country with an especially tumultuous political atmosphere – has seen an

---

[41] *https://www.bbc.co.uk/news/world-asia-58466939*.

increase in civilian access to social media. What could this rise in exposure to Internet content mean for the Afghan people, as well as people in other regions with significant levels of social unrest? With the rise of social media in the past decade, the power of terror groups has begun to transcend guns and bombs. In recent years, many have begun utilizing social media tools, such as YouTube, to convey threatening messages to the West, as well as to attract new recruits to join their cause. This strategy is a kind of psychological terrorism, which is aimed at intimidating Western forces, and at the same time, reeling in prospects by glorifying war in the name of freedom.

Why convey a message to the masses?

Apart from sending a warning to perceived oppressors, a deeper meaning behind emotional appeals used by terror groups might be a rallying cry to help round up more potential soldiers to join the cause. What specific elements do terrorists tend to use social media for? The message is often quite simple: Get off our land.

Making it look alluring

Such a referral to Westerners as oppressors by many groups identified as terrorists might bring to mind Samuel P. Huntington's world-renowned theory of the "clash" of civilizations. This theory holds that a clash of religious ideologies between East and West, particularly during the post-Cold War era, resulted in the majority of political conflicts thereafter.[42] If the "West" views a clash, clearly so do many groups within those cultures in the so-called "East."

Indeed, in terms of including an aspect of appeal for potential recruits, the montage of beautiful and horrid photos used by many of these groups in their propaganda represents a sort of gruesome miracle to be had in the name of freedom fighting. Such a concept of violence as precious serves as a clever method for attracting disgruntled civilians and any other interested viewers to jump on the bandwagon.

Overall, the classification of Western oppressors as enemy infidels can be compared to Huntington's argument regarding the East and West's portrayal of each other as "other" or somehow in the wrong, based on differing ideologies.

Using social media to attract and recruit

Over time, many terror groups have adopted social media as a strategy to spread their messages on an international scale. In combining classical, often aesthetic hymns and prayers with graphic footage of battles with Westerners, and memes of non-Western (in the case of white supremacists) forces, many such entities have succeeded in both setting the West on edge, as well as fueling the

---

[42] Huntington, Samuel P., *The Clash of Civilizations* (1996). Simon & Schuster.

contempt of many white nationalists in the West itself.[43] In the process, suicide for the sake of religion and violence for the alleged struggle for freedom has been placed on a pedestal. Likewise, violence in the name of nationalism has taken to the driver's seat in terms of white power propaganda. Perhaps most disturbingly, what unites cyber terrorism for the purposes of both extremism and nationalism, is the ability of anyone with an Internet connection to access such social media. In that sense, virtually anyone using these platforms can potentially fall victim to this perilous propaganda.

Next steps: Keep an eye on that friend request

Perhaps the strongest point to be made in terms of the use of social media to convey these ideals, is the fact that anyone with access to the Internet can potentially view videos of this sort. For once, one does not require an extensive education or even basic literacy. If extremist groups had hoped for a means of simultaneous popular recruitment and threats toward foreign military presence and perceived cultural dilution, YouTube has provided just that.

In fact, only recently have governments of countries such as the US noted social media as a breeding ground for many

---

[43] Fisher, M. "From Memes to Race War: How Extremists Use Popular Culture to Lure Recruits." *The Washington Post*, WP Company, April 30, 2021, https://www.washingtonpost.com/nation/2021/04/30/extremists-recruiting-culture-community/.

types of threats, from instigators of nationalism to romance scams.[44] Although companies can develop algorithms to watch for suspicious key words or potential malicious indicators, as in the areas of health care and social media content moderation, recruitment for cyber and geopolitical terrorists often strikes at the level of the individual user rather than on an organizational scale. Beginning from high school or even middle school, parents should pay attention to altered behavior in their children that seems to revolve around online connections pertaining to politics or violently counter-patriotic ideology.

Finally, this potential for extremism across social media and the corresponding need to watch our backs for insidious surveillance, echoes another rising point of worry in the realm of technology:

Drones that can fly over your home and monitor your activities.

---

[44] Fletcher, E. "Scams Starting on Social Media Proliferate in Early 2020." *Federal Trade Commission*, October 21, 2020, *https://www.ftc.gov/news-events/data-visualizations/data-spotlight/2020/10/scams-starting-social-media-proliferate-early-2020*.

CHAPTER SIX: DRONES – A SEARCH AND RESCUE SOLUTION OR THE ALL-SEEING EYE IN THE SKY?

Now that you are considering Facebook or Twitter as a potential venue for cyber crime – imagine who could be watching from the sky above.

Perhaps one of the most pressing topics in contemporary technology is the advancement and expanding use of unmanned aerial vehicles, more commonly referred to as UAVs or "drones."[45] The issues surrounding these pilotless aircraft involve questions of civilian privacy as well as security. Alongside the abundant controversy regarding American lethal UAVs in war zones abroad and their potential to harm innocents, comes the role of drones in non-martial activities, such as surveillance by state and local law enforcement within the US. With rapidly evolving technology in the US, particularly in California's Silicon Valley, it will be useful to examine the responses of state policymakers and society toward the privacy risks posed by law enforcement use of unmanned aircraft in the northern Californian city of San Jose.

---

[45] "DOD Developing Small, Unmanned Aerial System for Warfighters", *U.S. Department of Defense*. August 20, 2020. *https://www.defense.gov/News/News-Stories/Article/Article/2318919/dod-developing-small-unmanned-aerial-system-for-warfighters/*.

6: Drones – A search and rescue solution or the all-seeing eye in the sky?

John E. Murray, a 2016 Juris Doctorate candidate at Northeastern University School of Law, introduces the general premise of the privacy question posed by government and civilians alike when it comes to UAVs. As an attorney in training, Murray describes this widespread concern using the Fourth Amendment of the US Constitution's prohibition against unreasonable searches and seizures.[46] That is to say, he emphasizes the frequently disputed concept of how to precisely define a "search" when referring to law enforcement use of drones.[47] Until now, there has been a significant lack of legislation concerning police use of unmanned aerial surveillance, largely because of the relatively infrequent use of such aircraft by said agencies. The overall ruling by the Federal Aviation Administration (FAA) has declared rather vaguely that law enforcement is to employ the abilities of an unmanned aerial vehicle for purposes of civilian safety.[48] John Murray continues to further define what constitutes "civilian safety," in explaining that police often use UAVs to target criminals – such as drug dealers –

---

[46] U.S. CONST. amend. IV.
https://www.law.cornell.edu/constitution/fourth_amendment.

[47] Feeney, Matthew. "Does the 4th Amendment Prohibit Warrantless Drone Surveillance?" *CATO Institute*. March 24, 2021, https://www.cato.org/blog/does-4th-amendment-prohibit-warrantless-drone-surveillance.

[48] "Unmanned Aircraft Systems." *Federal Aviation Administration*. July 8, 2021. Web. October 24, 2014. https://www.faa.gov/uas/.

whose crimes may otherwise go undetected.[49] In this way, we may understand law enforcement surveillance via drones as a means of keeping criminals off the street, as well as defending the lives of innocent civilians in need of rescuing.[50] When considering their highly accelerated rate of disaster response, drones suddenly seem like an ideal – possibly even necessary – part of police operations.

So precisely what is it about the police adoption of UAVs that poses the question of privacy? Moving on to the state of California, Los Angeles political activist Hamid Khan advocates for the defense of civilian privacy rights via prohibition of law enforcement drone use. Referencing the Los Angeles police department's (LAPD) "long history of lies, brutality, and violence," Khan strongly opposes granting further power to the city's law enforcement.[51] In this way, we observe an example of societal concern over police abusing their access to high technology via intrusion

---

[49] Feeney, Matthew. "Does the 4th Amendment Prohibit Warrantless Drone Surveillance?" *CATO Institute*. March 24, 2021, *https://www.cato.org/blog/does-4th-amendment-prohibit-warrantless-drone-surveillance*.

[50] Dormehl, L. "Ingenious New Search and Rescue Drone finds People by Listening for Screams." *Digital Trends*, June 19, 2021, *https://www.digitaltrends.com/features/scream-scanning-search-and-rescue-drone/*.

[51] Ryan-Mosley, T., and Strong, J. "The Activist Dismantling Racist Police Algorithms." *MIT Technology Review*, June 5, 2020, *https://www.technologyreview.com/2020/06/05/1002709/the-activist-dismantling-racist-police-algorithms/*.

upon the private affairs of citizens. In an attempt to console such community unease, the LAPD has assured the public that it has no intention to militarize beyond UAV surveillance for rescue missions.[52] Nevertheless, Khan and his two dozen companions remain highly averse to police acquisition of unmanned aircraft, determined in their belief that advanced technology for surveillance would result in unjust privacy breaches. This criticism of police access to advanced technology for proposed enhanced security capability, indicates the problematic question of how to ensure the privacy of citizens from the very law enforcement sworn to protect them.

The technicalities of civilian privacy in UAV criminal investigation

Surely, one still cannot disregard the multifaceted ways in which UAVs can and already have facilitated daily human activity. After all, drones on the battlefield often mean fewer soldiers killed because of the decreased need for human pilots. On a more civilian note, UAVs could be adopted as package delivery vehicles, thus increasing delivery speed as well as decreasing the cost of travel by human postal workers. Furthermore, these unmanned aircraft are often smaller, less expensive to build and capable of longer flight than traditional aircraft.

---

[52] Reyes, R. "LAPD Drones Approved for Permanent Use." *Government Fleet*, September 19, 2019, https://www.government-fleet.com/340648/lapd-drones-approved-for-permanent-use.

6: Drones – A search and rescue solution or the all-seeing eye in the sky?

Where, then, does the problem arise? Let us return here to the question of when law enforcement using drones for surveillance transgresses the regulations set forth by the Federal Aviation Administration (FAA) to protect the privacy rights of US citizens, based upon the rights detailed in the Fourth Amendment.[53] Although citizens enjoy solid protection against warrantless government intrusions into their homes, the Fourth Amendment provides more lax restrictions upon government surveillance occurring in public places, including areas immediately outside the home, such as in yards or driveways. What might this mean for a civilian who was conducting illicit activities in his or her own backyard?

For a background on the controversy surrounding law enforcement drone use to observe civilians in our state of focus, let us examine the legal case of California v. Ciraolo. This case concerns a California resident named Dante Carlo Ciraolo who, following an anonymous tip off provided to Santa Clara police, was suspected of growing marijuana in his backyard.[54] In light of this evidence, the police felt entitled to survey Ciraolo's backyard, and just happened to discover that he was, indeed, cultivating

---

[53] The FAA is the US's largest transportation agency which regulates all civilian aviation activity in both domestic national territory and surrounding international waters.

[54] California v. Ciraolo. 476 U.S. 207. Supreme Court of the US. May 19, 1986. *LII / Legal Information Institute*. Cornell University Law School, October 25, 2014. Web. October 25, 2014. *https://www.law.cornell.edu/supremecourt/text/476/207*.

marijuana plants. However, because of the fact that the police had utilized a private airplane to search Ciraolo's yard from above, the defendant, as well as many dissenters of his criminality, argued that local government had acted without a warrant. Notwithstanding, the Supreme Court concluded the case by claiming that Ciraolo had no reasonable expectation of privacy, as any aircraft flying reasonably low over Ciraolo's yard could have easily seen the plants growing.

Now, what if law enforcement were to identify criminal activity using this sort of cutting-edge aerial technology, rather than regular manned aircraft? As civilians who are unaware of UAV surveillance cannot provide their permission to be searched, the use of advanced aerial technology such as drones to investigate their property may be deemed an invasion of privacy. Particularly regarding the sophisticated technology of some of these aircraft – such as thermal imaging capabilities used to detect heat involved in cooking or growing illicit drugs within a residence[55] – many individuals may grow suspicious of just how closely they are watched at home. Furthermore, as these UAVs are smaller and can therefore fly lower than manned aircraft, their line of sight into citizens' yards might provide police with better images than law enforcement could have ever gathered before incorporating the use of drones in a search. Therefore, the employment of

---

[55] Karlik, M. "Justices Critical of Continuous Home Video Surveillance without Police Warrant." *Colorado Politics*, April 23, 2021, *https://www.coloradopolitics.com/courts/justices-critical-of-continuous-home-video-surveillance-without-police-warrant/article_17db044e-96fa-11eb-91de-17535506acb9.html*.

6: Drones – A search and rescue solution or the all-seeing eye in the sky?

UAVs in a police surveillance mission might very well constitute an intrusive search, particularly if no warrant is obtained before the investigation.

San Jose as a case study

In California, as within any state or community, there are opposing views toward the proposition of law enforcement use of UAVs. Ever since 1999, when a San Jose police helicopter crashed killing both pilot and passenger, concern regarding police aircraft flying over the city has burgeoned. Because of a functional error, the helicopter began to spin in the air, forcing the officer, Desmond Casey, to crash-land the aircraft, fortunately in an area where no residents stood in the way.[56] In considering the issue of civilian safety as well as privacy implications, the use of law enforcement aerial technology seems doubly controversial. After all, if a manned aircraft can crash, what does this mean for a UAV that does not have a pilot on board to land it in a manner that avoids civilians? Still, as crucial as the question of ground security may be, not every crash can be predicted or prevented, and so we must venture on to the concern over another category of safety: civilian privacy.

The California legislature has addressed the potential privacy risk posed by drone surveillance with the proposed Bill AB-1327. Drafted by Assemblyman Jeff Gorell, this bill would require all public agencies in California –

---

[56] "SJPD Fallen Officers." *San Jose Police Department*, October 25, 1999. Web. December 4, 2014. *https://www.sjpd.org/about-us/inside-sjpd/department-information/fallen-officers*.

including law enforcement – to publicly announce their intention to deploy UAVs in national airspace within state jurisdiction and to obtain a warrant from those they intend to search.[57] At first glance, this bill might appear as a formidable protection of privacy rights of California citizens against any unwarranted police surveillance by drone. However, a closer look at the text of this legislation shows a fairly vague reference to "certain exceptions made for law enforcement in the instance that UAV use not be put toward tasks of criminal intelligence."[58] What exactly is meant by tasks of criminal intelligence, and do these exceptions imply free use of unmanned aircraft by police for civilian search? The bill proceeds to detail these exceptions as being instances in which innocent civilians are at risk of "great bodily harm."[59] Most importantly, what

---

[57] Gorell, J. "Bill Text AB-1327 Unmanned Aircraft Systems." *California Legislative Information.* February 22, 2013. Web. July 8, 2021.
https://leginfo.legislature.ca.gov/faces/billNavClient.xhtml?bill_id=201320140AB1327.

[58] Gorell, J. "Bill Text AB-1327 Unmanned Aircraft Systems." *California Legislative Information.* February 22, 2013. Web. July 8, 2021.
https://leginfo.legislature.ca.gov/faces/billNavClient.xhtml?bill_id=201320140AB1327.

[59] Gorell, J. "Bill Text AB-1327 Unmanned Aircraft Systems." *California Legislative Information.* February 22, 2013. Web. July 8, 2021.
https://leginfo.legislature.ca.gov/faces/billNavClient.xhtml?bill_id=201320140AB1327.

do these citizens have to have done or be suspected to have done in order to warrant such an investigation, if not criminal activities?

This rather open-ended wording of Bill AB-1327 has, understandably, provoked high levels of civilian activism both in favor of, and opposed to, the passing of this legislation. Because of its location in the heart of the Silicon Valley, the Santa Clary County city of San Jose has experienced an especially tumultuous outcry from both sides in response to this new policy that essentially combines privacy rights with the use of advanced technology.

Those opposed

This struggle to strike a balance between privacy protection and the benefit of advanced surveillance tools for enhanced security, is a multifaceted issue, which is not solely in response to the still ongoing controversy surrounding this bill. Indeed, in a region such as the Silicon Valley, the advantage of increasingly sophisticated technology must necessarily be weighed against aspects of societal rights. In particular, skeptics of Bill AB-1327 emphasize its elusive section stating that police only require a warrant to search provided the citizen in question has a "reasonable expectation of privacy."[60] Once again, we encounter the

---

[60] Gorell, J. "Bill Text AB-1327 Unmanned Aircraft Systems." *California Legislative Information.* February 22, 2013. Web. July 8, 2021.
https://leginfo.legislature.ca.gov/faces/billNavClient.xhtml?bill_id=201320140AB1327.

familiar issue touched upon in the California v. Ciraolo case concerning what factors denote a reasonable expectation of privacy?

Despite Assemblyman Gorell's emphasis on the need for a search warrant,[61] many critics of this new bill argue that it fails to take serious enough measures to regulate exactly when law enforcement should require a warrant. The suspicions spurred by this grey area continue northward from Hamid Khan's protests in Los Angeles. In the case of San Jose, Nicole Ozer, the Technology and Civil Liberties Director for the American Civil Liberties Union of California, expresses concern over the logistics of UAV incorporation into police activities.[62] Although the San Jose Police Department has promised to use drones solely for purposes of bomb detection, Ozer describes this statement as suggestive of a slippery slope, in which the line between serious threat and other reasons for police surveillance may blur. She further explains her opinion in stating the possibility of a police force using its advanced aerial technology to monitor political protests and spy on minority communities under the guise of threat prevention.

---

[61] Gorell, J. "Bill Text AB-1327 Unmanned Aircraft Systems." *California Legislative Information.* February 22, 2013. Web. July 8, 2021. *https://leginfo.legislature.ca.gov/faces/billNavClient.xhtml?bill_id=201320140AB1327*.

[62] Abraham, R. "Inside the ACLU's Nationwide Campaign to Curb Police Surveillance." *The Verge*, June 14, 2017, *https://www.theverge.com/2017/6/14/15795056/aclu-police-surveillance-curb-campaign-nationwide*.

6: Drones – A search and rescue solution or the all-seeing eye in the sky?

Particularly following the discovery that the National Security Agency had been spying on US citizens in recent years, the public are becoming increasingly educated on issues of government surveillance and how law enforcement may abuse technology to conduct espionage on civilians. Indeed, the fact that San Jose City Council has simply agreed to sponsor its police department's drone purchase has inflamed many citizens opposed to government UAV surveillance without what these citizens feel is a clear usage policy.[63] Although the drone that was purchased by the San Jose Police Department is relatively rudimentary, three-foot-wide Century Neo 660 V2, [64] produced by the local Century Helicopter Products, critics remain skeptical. After all, this makes San Jose the first city in the Bay Area region whose law enforcement now possesses a drone.[65]

---

[63] Salonga, R. "San Jose Police Drone Inflames Surveillance." *San Jose Mercury News*. August 1, 2014. Web. July 8, 2021. *https://www.mercurynews.com/2014/07/31/san-jose-police-drone-inflames-surveillance-state-rumblings/*.

[64] Salonga, R. "San Jose Police Drone Inflames Surveillance." *San Jose Mercury News*. August 1, 2014. Web. July 8, 2021. *https://www.mercurynews.com/2014/07/31/san-jose-police-drone-inflames-surveillance-state-rumblings/*.

[65] Marzullo, K. "Report: SJPD Is First Local Agency to Get a Drone." *ABC7 San Francisco*. ABC Inc., July 8, 2021. Web. October 26, 2014. *https://abc7news.com/san-jose-police-sjpd-drones-drone/229565/*.

6: Drones – A search and rescue solution or the all-seeing eye in the sky?

In the wake of the drone purchase by San Jose police, Charlotte Casey from the San Jose Peace and Justice Center – a South Bay activist group concerned with regional social justice – issued a petition of more than 1,600 signatures demanding that the drone be returned. Casey claims that a UAV constitutes "distancing technology" in a society where the community strives for an honest relationship with its police force, particularly when considering that the drone in question comes equipped with a GoPro camera. From Casey's stated concerns, the general opinion of those opposed to law enforcement drone use seems to include the very existence of such technology at the city police department. Such aversion illustrates the distrust held by these individuals regarding the officers' word to never deploy a UAV for espionage purposes.

Aware of this immense anxiety among certain resident groups, San Jose police spokesman, Officer Albert Morales, has adamantly assured the public that although the actual flight of drones for police activities still awaits FAA approval, their deployment would represent immense advantages toward the effectiveness of law enforcement protection of civilians.[66] Morales' cautious attitude in embracing the use of UAVs while simultaneously acknowledging the benefits of doing so, suggests a reaction toward the widespread opposition to the role of drones in police activity. Law enforcement members understand the

---

[66] Marzullo, K. "Report: SJPD Is First Local Agency to Get a Drone." *ABC7 San Francisco*. ABC Inc., July 8, 2021. Web. October 26, 2014. *https://abc7news.com/san-jose-police-sjpd-drones-drone/229565/*.

need to tread carefully in matters of technology and surveillance, particularly when seeking to appease a dubious public.

It was in response to such abundant opposition toward law enforcement deployment of UAVs that the California State Senate originally approved the Bill AB-1327. In fact, the bill's emphasis on advocating privacy rights in the form of public announcement, as well as a warrant required for all police searches involving unmanned aircraft, highlights the opposing side's Fourth Amendment concerns in light of government access to rapidly advancing technology. Thus far, the fate of this bill depended upon the decision of California Governor Jerry Brown as to whether to pass or veto. As it happened, the Governor vetoed AB-1327. Assemblyman Jeff Gorell, author of the bill, has expressed frustration over what he views as Governor Brown supporting law enforcement over the privacy interests of Californians. Moreover, Gorell even goes on to warn of a potential override by the California State Legislature banning drone use in the entire state, to help ensure the security and peace of mind of the California community.[67] Indeed, it stands to reason that the governor's refusal to pass this bill protecting privacy rights would understandably anger as well as concern critics of police access to UAVs.

---

[67] Willon, P., and Mason, M. (2014) "Governor Vetoes Bill That Would Have Limited Police Use of Drones." *Los Angeles Times*, *https://www.latimes.com/local/political/la-me-ln-governor-vetoes-bill-to-limit-police-use-of-drones-20140928-story.html*.

Essentially, the governor's response to this bill indicates his resolve that, although skepticism over the security of police drone surveillance is reasonable, this particular bill goes too far in its call for a warrant before every search, save for direst emergencies such as bomb threats. This line of thinking represents Governor Brown's split opinion over those citizens opposed to police use of UAV and law enforcement's view favoring the deployment of such technology. In the end, despite the misgivings of many critics, he appears to lean more toward the path to enhanced police surveillance via these unmanned aircraft.

Those in favor

In light of this seemingly widespread concern over unmanned aircraft integration into police operations, what exactly are the aspects of law enforcement utilizing UAVs that individuals such as Governor Brown and members of the San Jose police force find so enticing? For one, the police department has stressed the Century Neo 66's ability to rescue a bomb squad technician, provided its advanced accuracy in detecting where a human may safely venture in the case of a bomb threat. Here we encounter the other side of the coin – the invaluable advantages of employing such technology to save lives. Would not the potential to safeguard human life outweigh the possible occasional instance of over-surveillance provided by drones?

In the police department's drone application drafted to the Bay Area Urban Areas Security Initiative of the Department of Homeland Security, the capability of a UAV to improve the overall response time of a bomb squad is highly emphasized.[68] In fact, upon the revelation that the San Jose police had purchased a drone, the department issued a statement highlighting the exact ways in which unmanned aircraft could significantly enhance the ability of law enforcement to protect citizens from hazards. The statement provides the following details regarding hazard type and location:

> *"We are confident that this technology can improve certain police operational efficiencies and help enhance public and officer safety in specific critical incidents.*
>
> *SJPD [San Jose Police Department] strives to explore new technology that can help our department protect the community more effectively and efficiently."*[69]
>
> *"The UAS [unmanned aerial system] can be flown over a device to obtain images that would assist the bomb technicians. Another possible use would be for situations that threaten public safety. These could include dangers such as active shooters, hostage taking,*

---

[68] "San Jose Police Apologizes for Secrecy Surrounding Purchase of Drone." *CBS San Francisco.* August 5, 2014. Web. July 8, 2021. *https://sanfrancisco.cbslocal.com/2014/08/05/san-jose-police-apologizes-for-secrecy-surrounding-purchase-of-drone-crime-privacy-surveillance-drones-unmanned-aerial-vehicle/.*

[69] "San Jose Police Apologizes for Secrecy Surrounding Purchase of Drone." *CBS San Francisco.* August 5, 2014. Web. July 8, 2021. *https://sanfrancisco.cbslocal.com/2014/08/05/san-jose-police-apologizes-for-secrecy-surrounding-purchase-of-drone-crime-privacy-surveillance-drones-unmanned-aerial-vehicle/.*

or other such tactical situations where lives might be in immediate danger."[70]

In order to provide a closer look at the multifaceted role of a police UAV, employees of the San Jose Police Department were kind enough to elaborate upon what they and many of their colleagues view as the essential benefits of deploying drones in place of human officers. During personal interviews with the author, these officers have remained anonymous while offering the following insight:

1. In your opinion, what could be the greatest advantage to using a drone during SJPD operations?

Officer #1: *"The Unmanned Aerial System (UAS) would allow us to use the UAS to approach potential explosive devices and gain tactical information when a person's life was in danger (such as active shooter, hostage/barricade situation) without exposing a police officer to that peril.*"[71]

Here, Officer #1 captures the unique capability of a UAV to provide more accurate data than could be

---

[70] "San Jose Police Apologizes for Secrecy Surrounding Purchase of Drone." *CBS San Francisco.* August 5, 2014. Web. July 8, 2021. *https://sanfrancisco.cbslocal.com/2014/08/05/san-jose-police-apologizes-for-secrecy-surrounding-purchase-of-drone-crime-privacy-surveillance-drones-unmanned-aerial-vehicle/.*

[71] "SJPD Use of Drones." Telephone interview. December 3, 2014. Telephone interview conducted as part of a graduate symposium on law enforcement use of UAVs.

otherwise collected. This sort of data would allow officers to approach a perilous situation with better foresight, therefore allowing them a higher chance of survival should matters grow dire.

Officer #2: *"The greatest advantage is the ability to look into and onto areas we can't see otherwise, during tactical situations, such as hostage barricades, bomb callouts, active shooter scenarios, and disasters such as floods and poisonous chemical spills. The ability to see and gather more information greatly enhances the safety of our officers by eliminating some of the unknown element and can help locate and isolate armed perpetrators and identify hazards. In the case of a bomb callout, the drone can search for secondary devices before officers enter the scene, it can look for persons with detonators, it can help determine what kind of device it is, and look for other dangers such as snipers or traps."[72]*

Following up on the general information provided by the San Jose Police Department on the advantages of aerial technology as well as the viewpoint offered by Officer #1 regarding risk minimization to human life, Officer #2 gives a more detailed perspective as to exactly how the advanced technology of a UAV could be used in a vast array of disaster situations. In describing a drone's ability to detect a bomb or shooter threat before a human officer could,

---

[72] "SJPD Use of Drones." Telephone interview. December 3, 2014. Telephone interview conducted as part of a graduate symposium on law enforcement use of UAVs.

6: Drones – A search and rescue solution or the all-seeing eye in the sky?

Officer #2 highlights the unique use a drone could offer its police squad. In particular, he discusses the "unknown element," frequently the most dangerous aspect of a task when not even the expert team knows what to expect. Such sophisticated threat assessment capabilities could be utilized in rescue missions, as well as hazards, in order to minimize damage to both civilians as well as police officers.

2. In your opinion, what could be the most prominent disadvantage to using a drone during SJPD operations?

Officer #1: *"The disadvantages could be the perceived thought by the public that the police are using it to watch them. This is why we are carefully crafting policy and conducting community outreach to alleviate this misperception."*

Officer #1's reference to the public perception of and subsequent anxiety about UAV-related espionage, speaks to the main concerns emphasized in the privacy rights arguments made by organizations such as the American Civil Liberties Union (ACLU) and the San Jose Peace and Justice Center. Officer #1 elaborates upon this privacy obstacle by highlighting the police department's goal to draft policy relating specifically to public fear over intrusive surveillance.

Officer #2: *"The only disadvantage I can think of is that without the proper policies and procedures, the drone could be an invasion of privacy and possibly a violation of the Fourth Amendment. It is important that these* *policies are vetted by the community and kept in check through constant scrutiny.* "[73]

Here, Officer #2 voices his only concern regarding police use of drones, an opinion that seems to largely echo that of the civil rights activist groups when it comes to privacy rights. Hearkening back to the original purpose for drafting Bill AB-1327, the issue of Fourth Amendment violations arises with the adoption of a police UAV. Overall, Officer #2 stresses the necessity for public consent where such privacy obstacles come into play.

3. What sort of general consensus, if any, have you observed from the majority of your fellow officers regarding drone deployment during police operations?

Officer #2: *"We have never deployed the UAS. The majority of police officers feel that it would be very beneficial and assist them with safety concerns. We are only discussing that the UAS be used in the very limited situations listed above. We do not have any interest in using the UAS for surveillance purposes."* [74]

In a response that reflects the idea behind Officer Morales' statement, Officer #1 insists that not only will the police force never use a drone to spy on civilians, but they will not

---

[73] "SJPD Use of Drones." Telephone interview. December 3, 2014. Telephone interview conducted as part of a graduate symposium on law enforcement use of UAVs.

[74] "SJPD Use of Drones." Telephone interview. December 3, 2014. Telephone interview conducted as part of a graduate symposium on law enforcement use of UAVs.

even utilize its capabilities at all, save in dire situations such as a bomb threat. This reassurance as to the minimal hypothetical use of the UAV, as well as the confirmation that the department has never deployed the vehicle, both serve to uphold the officers' promise to the public that the drone will not be used without full agreement of the people.

Officer #2: *"The general consensus that I have observed among my colleagues and coworkers is that it [the UAV] is a very useful tool than can save lives. It is a tool to increase effectiveness. It does need to have specific policies and guidelines that must be enforced and guarded."*[75]

Despite Officer #2's more personal response to the previous question regarding possible drawbacks in using drones for disaster and threat situations, and in line with Officer #1's outlook on the extreme security potential of UAVs, the general opinion of the San Jose Police Department seems to value benefit over risk. Understandably, the police who put their lives on the line in their mission to rescue and protect civilians would want to kill two birds with one stone. That is to say, any opportunity to improve civilian security while simultaneously mitigating the danger posed when an officer approaches a hazardous environment, is too promising a prospect to turn down.

---

[75] "SJPD Use of Drones." Telephone interview. December 3, 2014. Telephone interview conducted as part of a graduate symposium on law enforcement use of UAVs.

6: Drones – A search and rescue solution or the all-seeing eye in the sky?

Regardless of the invaluable benefits presented by advanced detection tools in police activities, such tools will be difficult to employ without the clear approval of the public. Therefore, the specific correlation between law enforcement drone use and the anxiety over privacy demonstrated by the citizens of San Jose, must be solidly handled.

Calming the public outcry

Now what do the experts have to say on the technical aspect of just how these aerial vehicles could be used for purposes of security and espionage? To add to the aforementioned description of crucial opportunities offered by law enforcement UAVs, Gary Richard Lane, a specialist in aviation law and licensed pilot based out of New Hampshire, addresses the fear of privacy violations that often seems to outweigh the pros of drone deployment in rescue missions.

> *"This isn't about police wanting to have a drone hovering in your backyard watching you in your living room or swimming pool. Besides, there are much more effective ways for cops to snoop on you than with a drone (Lane)."*[76]

---

[76] Johnson, M. Alex. "Drones Are Ready to Save Lives, But U.S. Regulations Keep Them on the Ground." *NBC News*. May 23, 2014. Web. July 8, 2021. *https://www.nbcnews.com/news/us-news/drones-are-ready-save-lives-u-s-regulations-keep-them-n113611*.

Although Lane stresses a number of valid points in his assessment of privacy concerns posed by the role of drones in law enforcement activity, this worry over such civil rights violation still constitutes a cry loud enough to have kept police drones on the ground thus far, as depicted by the current situation in San Jose. For this reason in particular, state government officials such as Governor Brown would do well to attend to the worries of their citizens, if they support a more capable police force equipped with UAVs.

In any instance, the case of drone deployment by San Jose officers is far from over, its continuation was evident in the public hearing on December 6, 2014. Citizens from around the Bay Area were welcome to attend and voice any concerns during a question-and-answer session following the meeting. Such an opportunity for public participation in police affairs that may affect civilian privacy, constitutes a crucial step in addressing the main obstacle that has, up until now, stood between the San Jose Police Department and drone use. Provided the opportunities for enhanced safety measures detailed by the two officer interviews as well as Mr. Lane, the subject of police UAVs deserves critical discussion, preferably the kind in which the people of the San Jose region take the largest part.

Balancing technological surveillance measures with public interest

Although the presentation of the benefits of police drone use highlighted above may seem sedate compared to the protests of civil liberty advocate groups pressing for privacy rights, the advantages speak for themselves. There is an unmistakable opportunity to better ensure the safety of officers and threat response teams by employing drones in place of human law enforcement, when appropriate. After all, unmanned aircraft are equipped with threat detection sensors superior to those of humans. Similarly, a drone being blown up by a bomb is favorable to a human in its place, not unlike the purpose served by military UAVs on battlefields abroad. Indeed, provided the San Jose community's location in Silicon Valley, the evolving public opinion and reactions toward law enforcement activities offers a case study most relevant to the theme of technology and society.

Needless to say, the issue of civilian privacy rights outlined in the United States' Constitution also deserves solid consideration in the face of police drone deployment. As matters now stand, the dispute appears to be between law enforcement officials who wish to utilize unmanned aerial vehicles toward improvement of operational efficiency versus those factions of public society who remain highly skeptical of potential civil rights breaches implicated by the use of such advanced technology. In reality, neither the dubiousness over privacy protection nor the potential of UAVs to save many lives can be ignored. In light of the pressing anxiety because of a lack of warrant as well as lucidity regarding the time and location of police drone deployment, the most appropriate solution would likely involve an entirely transparent description of exactly how and when the San Jose Police Department plan to utilize unmanned aircraft for rescue and security purposes, and that the force keeps its word regarding the reservation of its UAV for only the most critical situations.

However, in order to strike this balance between UAV incorporation by law enforcement and community satisfaction within California, state government must first heed the privacy concerns of its constituents. If the accounts given by the two police officers offer any indication, the San Jose Police Department has already made it a priority to quell public fear over the unmanned aerial technology that could revolutionize the security of both civilians and law enforcement for the next generation.

Next steps: Securing the wireless target

That said, with all that background about steps taken to address concerns over civilian privacy – could criminal hackers have the ability to potentially hack into a search and rescue drone? Although a quick-moving UAV never exposed to physical access by attackers would, in theory, avoid most security risks evident in tampering with the device itself, it only takes one to cause a catastrophic domino effect. All it might take is one attacker in the right place at the right time to succeed at sniffing a drone's connectivity to certain ports, such as for file transfer or telnet for wireless connection. For example, imagine one search and rescue or even a simple Amazon delivery drone falls into the hands of an attacker who then analyzes that drone's data to learn and control the programming of all interconnected drones. Such interference runs the risk of setting off a distributed denial-of-service (DDoS) attack that diverts a dangerous rescue mission or prevents hundreds of customers from receiving their packages. Despite the potential implementation costs, drone manufacturers as well as individual users could benefit from investing in reliable VPNs for all drone communications, as well as continuous assessment of wireless network activity so as to avoid the risk of unused port connections that leave UAVs vulnerable to unauthorized connections from potential attackers.

Recent developments

While San Jose represented a focal point for major developments in drone deployment back in 2014 and 2015, the use of military and police drones has since expanded into the greater Bay Area, including Hayward and Oakland. Indeed, in early 2022, the City of Hayward confirmed surveillance policy for police use of drones.[77] Furthermore, the Oakland Police Department has recently announced the intended launch of an entire program dedicated to drone deployment. [78] On both fronts, surveillance regulators continue with close prioritization of civilian privacy rights amidst these new developments.

---

[77] Cabral, A. (2022). "Hayward Sets Police Drone Use Guidelines". *The Mercury News*, https://www.mercurynews.com/2022/01/26/hayward-sets-police-drone-use-guidelines/.

[78] Kelly, G. (2022). "Oakland Police to Launch Drone Program." *East Bay Times*, https://www.eastbaytimes.com/2022/03/14/oakland-police-to-launch-drone-program/.

CHAPTER SEVEN: TYING IT ALL TOGETHER

At this point we've discussed all manner of worrisome situations thanks to the rapid-fire development of technology. Starting off with the COVID-19 pandemic, we saw how attackers continue to play off people's fear to trick them into engaging with the criminal hacker. Then, continuing along this pathway of emotional response, we observe just how susceptible many social media users are to online content – both in terms of the lay Facebook browser as well as the content moderation employee staring at graphic material day in and day out, all because unseemly types choose to use the platform for nefarious purposes. Moving further, we consider how terrorism – a frightening phenomenon that has always prevailed by playing upon the negative emotions of youth and the disenfranchised – has reached a plethora of new potential recruits once taking to social media platforms. Finally, we move from cyberspace to that eye in the sky – the computerized drone that could be flying right above your home, its artificial intelligence only increasing as technology progresses each day.

By now, you might be thinking that attackers lurk at every corner – in the sky, on your computer, and even spying on your voting preferences. Indeed, the past decade alone has witnessed scores of cyber attacks involving phishing, terror recruitment, elections meddling, and increasing concern over technology used for police surveillance. Especially in the wake of COVID-19 and heightened anxiety over how terrorism and law enforcement continue to adapt to new technology, these areas present a starting point for examining the massive scope of the threat landscape we face today.

After tearing through the threats of phishing, cyber terrorism, and even drone hacking, let's take a breath. Is all this information and stress even worth it, or are we getting ahead of ourselves by delving so deep into the dark side of the online world? Indeed, perhaps the title of this book even seems to poke fun at those who worry about the safety of their health clinic, their email account, their Facebook profile, or their remote-control garage door opener. Rest assured, however, that such paranoia is hardly unwarranted. An interconnected reality opens up just as many avenues for attack as for innovation. In the face of brilliant attackers eager for the moment to strike, watching your back is your best bet. This book aims to provide a starting point as far as mitigating steps to help avoid those threats.

APPENDIX A: BIBLIOGRAPHY

Abraham, R. "Inside the ACLU's Nationwide Campaign to Curb Police Surveillance." The Verge, June 14, 2017, *https://www.theverge.com/2017/6/14/15795056/aclu-police-surveillance-curb-campaign-nationwide*.

Akpan, N. "Ransomware and Data Breaches Linked to Uptick in Fatal Heart Attacks." *PBS*, October 24, 2019, *https://www.pbs.org/newshour/science/ransomware-and-other-data-breaches-linked-to-uptick-in-fatal-heart-attacks*.

Allodi, L., et al. "The Need for New Antiphishing Measures Against Spear-Phishing Attacks." *IEEE Security & Privacy*, vol. 18, no. 2, September 20, 2019, pp. 23–34, doi:10.1109/msec.2019.2940952. APWG, *https://apwg.org/trendsreports/*.

Bann Lap, L., et al. "Trusted Security Policies for Tackling Advanced Persistent Threat via Spear Phishing in BYOD Environment." *Procedia Computer Science*, vol. 72, 2015, pp. 129–36, doi:10.1016/j.procs.2015.12.113.

Baryshevtsev, M. (05/2020). "Persuasive Appeals Predict Credibility Judgments of Phishing Messages." *Cyberpsychology, Behavior, and Social Networking* (2152–715), 23 (5), 297.

Baxter, G., & Sommerville, I. (2011). "Socio-technical systems: From design methods to systems engineering, Interacting with computers," 23 (1), 4–17.

BBC News. (2021). "The Taliban embrace social media: 'We too want to change perceptions,'" *https://www.bbc.co.uk/news/world-asia-58466939*.

Bossetta, M. (2018). "The Weaponization of Social Media: Spear Phishing and Cyber Attacks on Democracy," *Journal of International Affairs*, bind 2018 Special Issue, vol. 71, nr. 2, 6, s. 97–106.

Botacin, M., et al. "We Need to Talk about Antiviruses: Challenges & Pitfalls of AV Evaluations." *Computers & Security*, vol. 95, August 2020, p. 101859, doi:10.1016/j.cose.2020.101859.

Brumley, K. "COVID-19 Scam Alerts." *Cybercrime Support Network*, Guidestar Silver Transparency, June 21, 2021, *https://cybercrimesupport.org/covid-19-scam-alerts/*.

Cabral, A. (2022). "Hayward Sets Police Drone Use Guidelines". *The Mercury News*, *https://www.mercurynews.com/2022/01/26/hayward-sets-police-drone-use-guidelines/*.

California v. Ciraolo. 84–1513. "Supreme Court of the US." 1985. *The Oyez Project at IIT Chicago-Kent College of Law*. Web. October 25, 2014, *https://www.oyez.org/cases/1985/84-1513*.

California v. Ciraolo. 476 US 207. Supreme Court of the US. May 19, 1986. *LII / Legal Information Institute*. Cornell University Law School, October 25, 2014. Web.

October 25, 2014,
https://www.law.cornell.edu/supremecourt/text/476/207.

CBS San Francisco, "San Jose Police Apologizes for Secrecy Surrounding Purchase of Drone." August 5, 2014. Web. July 8, 2021, *https://sanfrancisco.cbslocal.com/2014/08/05/san-jose-police-apologizes-for-secrecy-surrounding-purchase-of-drone-crime-privacy-surveillance-drones-unmanned-aerial-vehicle/.*

Chapman, P. (04/2020). "Are your IT staff ready for the pandemic-driven insider threat?" *Network Security* (1353–4858), 2020 (4), 8.

Chaykowski, K. "Facebook Publishes Internal Content Moderation Guidelines For The First Time." *Forbes Magazine*, April 25, 2018, *https://www.forbes.com/sites/kathleenchaykowski/2018/04/24/facebook-publishes-internal-content-moderation-guidelines-for-the-first-time/?sh=ce373cf71e7d.*

Chetty, N., Alathur S. "An Architecture for Digital Hate Content Reduction with Mobile Edge Computing." *Digital Communications and Networks*, vol. 6, no. 2, 2020, pp. 217–22, doi:10.1016/j.dcan.2019.05.004.

Chowdhury H., et al. "Time Pressure in Human Cybersecurity Behavior: Theoretical Framework and Countermeasures." *Computers & Security*, vol. 97, October 2020, p. 101963., doi:10.1016/j.cose.2020.101963.

Criddle, C. "Facebook Moderator: 'Every Day Was a Nightmare.'" BBC News, May 12, 2021, *https://www.bbc.co.uk/news/technology-57088382.*

Christopher, N. "Facebook's Fake News Clean-up Hits Language Barrier." *The Economic Times*, April 13, 2018.

CNN, October 28, 2020, "2016 Presidential Campaign Hacking Fast Facts," *https://edition.cnn.com/2016/12/26/us/2016-presidential-campaign-hacking-fast-facts/index.html*.

"Cofense Launches Free Resource Center and Searchable Database Highlighting the Latest Phishing Attacks that Bypass Email Security Technologies." (June 9, 2020). *UWIRE Text*, p. 1.

"Cognizant Turning Away From Vile Online Content Control. (October 31, 2019)." *International Business Times* [US ed.], p. NA. Retrieved from *https://techxplore.com/news/2019-10-cognizant-vile-online-content.html*.

"Common, F. M. Fear the Reaper: how content moderation rules are enforced on social media." Pages 126–52. Received September 30, 2019. Accepted January 30, 2020. Published online: March 6, 2020.

Congressional Research Service. *Drones in Domestic Surveillance Operations: Fourth Amendment Implications and Legislative Responses*. By Richard M. Thompson. 113nd Cong., 1st sess. Cong. Rept. 7–5700. N.p.: April 2013. Web. October 4, 2014, *https://www.defense-aerospace.com/articles-view/reports/2/138277/concerns-about-domestic-drone-surveillance.html*.

"Content moderation is hard, but there's a new approach and it's fueled by Spectrum Labs." (January 28, 2020.) *Marketwired*, p. NA. Retrieved from *https://www.globenewswire.com/news-*

release/2020/01/28/1975839/0/en/Updated-Content-moderation-is-hard-but-there-s-a-new-approach-and-it-s-fueled-by-Spectrum-Labs.html.

"Daly, J. History of Email In The Federal Government." *Technology Solutions That Drive Government*, CDW, November 27, 2019, *https://fedtechmagazine.com/article/2013/07/brief-history-email-federal-government*.

Davis, C. "Anti-Spam Message Headers – Office 365." *Anti-Spam Message Headers – Office 365 | Microsoft Docs*, GitHub, Inc., June 16, 2020, *https://docs.microsoft.com/en-us/microsoft-365/security/?view=o365-worldwide*.

"DOD Developing Small, Unmanned Aerial System for Warfighters", *U.S. Department of Defense*. August 20, 2020, *https://www.defense.gov/News/News-Stories/Article/Article/2318919/dod-developing-small-unmanned-aerial-system-for-warfighters/*.

Dormehl, L. "Ingenious New Search and Rescue Drone finds People by Listening for Screams." *Digital Trends*, June 19, 2021, *https://www.digitaltrends.com/features/scream-scanning-search-and-rescue-drone/*.

Dukarm, Christopher; Dill, Richard; Reith, Mark. (2019). "Proceedings of the European Conference on Cyber Warfare & Security." ACPIL, p172–7.

Earl, J. "Google's Top Search Result for 'Final Election Numbers' Leads to Fake News Site." *CBS News*, CBS Interactive, November 15, 2016, *https://www.cbsnews.com/news/googles-top-search-result-for-final-election-numbers-leads-to-fake-news-site/*.

Editorial: "Drone Law Is Urgent for California. Justice and Public Safety." *San Jose Mercury News*, June 9, 2014. Web. October 25, 2014.

"Facebook Expands AI Translation Tools To Increase Its Global Content Moderation." *New York Law Journal* (0028–7326), 260 (53), 5.

Feeney, Matthew. "Does the 4th Amendment Prohibit Warrantless Drone Surveillance?" *CATO Institute*. March 24, 2021, *https://www.cato.org/blog/does-4th-amendment-prohibit-warrantless-drone-surveillance*.

Fisher, M. "From Memes to Race War: How Extremists Use Popular Culture to Lure Recruits." *The Washington Post*, WP Company, April 30, 2021, *https://www.washingtonpost.com/nation/2021/04/30/extremists-recruiting-culture-community/*.

Fletcher, E. "Scams Starting on Social Media Proliferate in Early 2020." *Federal Trade Commission*, October 21, 2020, *https://www.ftc.gov/news-events/data-visualizations/data-spotlight/2020/10/scams-starting-social-media-proliferate-early-2020*.

Franscella, J. Anomali "Threat Research Team Identifies Widespread Credential Theft Campaign Aimed at US and International Government Agency Procurement Services." (December 12, 2019). *Marketwired*, p. NA.

Gioe V. "Cyber Operations and Useful Fools: the Approach of Russian Hybrid Intelligence." *Intelligence and National Security*, vol. 33, no. 7, December 28, 2018, pp. 954–73, doi:10.1080/02684527.2018.1479345.

Gorell, J. Bill Text AB-1327 Unmanned Aircraft Systems. *California Legislative Information*. February 22, 2013. Web. October 25, 2014, *https://leginfo.legislature.ca.gov/faces/billNavClient.xhtml?bill_id=201320140AB1327*.

Grauer, Y. "Why Is the Healthcare Industry Still so Bad at Cybersecurity?" *ArsTechnica, WIRED Media Group*, February 9, 2020, *https://arstechnica.com/information-technology/2020/02/why-is-the-healthcare-industry-still-so-bad-at-cybersecurity/*.

Greenstein, Nicole. "Privacy and the Law: How the Supreme Court Defines a Controversial Right." *US Privacy and the Law*. Time Inc., July 31, 2013. Web. October 29, 2014, *https://nation.time.com/2013/08/01/privacy-and-the-law-how-the-supreme-court-defines-a-controversial-right/slide/thermal-imaging-devices/*.

Gregorio, D. "Democratising Online Content Moderation: A Constitutional Framework." *Computer Law & Security Review*, vol. 36, 2020, p. 105374, doi:10.1016/j.clsr.2019.105374.

Haro, F. (May 12, 2020). "FBI warns against COVID-19 phishing scams." *UWIRE Text*, p. 1.

Heaton, B. "California Legislature Approves Drone Regulation." *Solutions for State and Local Government Technology*. N.p., August 28, 2014. Web. October 26,

2014. *https://www.govtech.com/public-safety/california-legislature-approves-drone-regulation.html*.

Hewitt, R. "WebPurify Says AI Helping Hand for Image Moderation, but Human Intervention Is Needed." *Business Wire*, January 22, 2019, *https://www.businesswire.com/news/home/201901220050 23/en/*.

Hoffman, B. *Inside Terrorism*. (New York: Columbia University Press, 2006), p. 40.

Hohenstein, J., Malte J. "AI as a Moral Crumple Zone: The Effects of AI Mediated Communication on Attribution and Trust." *Computers in Human Behavior*, vol. 106, 2020, p.106190, doi:10.1016/j.chb.2019.106190.

Huntington, S. P. "The Clash of Civilizations?" *Foreign Affairs*. Council on Foreign Relations, 1993. Web. April 28, 2014.

Johnson, M. Alex. "Drones Are Ready to Save Lives, But U.S. Regulations Keep Them on the Ground." *NBC News*. May 23, 2014. Web. July 8, 2021, *https://www.nbcnews.com/news/us-news/drones-are-ready-save-lives-u-s-regulations-keep-them-n113611*.

Karlik, M. "Justices Critical of Continuous Home Video Surveillance without Police Warrant." Colorado Politics, April 23, 2021, *https://www.coloradopolitics.com/courts/justices-critical-of-continuous-home-video-surveillance-without-police-warrant/article_17db044e-96fa-11eb-91de-17535506acb9.html*.

Kelly, G. (2022). "Oakland Police to Launch Drone Program." *East Bay Times*, *https://www.eastbaytimes.com/2022/03/14/oakland-police-to-launch-drone-program/*.

Kumar, M., Hanumanthappa, M., Kumar, S., T. V.. A "Countermeasure Technique for Email Spoofing. International Journal of Advanced Research in Computer Science," Jan/Feb 2013, Vol. 4, Issue 1, p128–33, 6p.

Langvardt, K. (June 2018). "Regulating Online Content Moderation." *Georgetown Law Journal*, 106 (5), 1353+.

Lardieri, Alexa. "Cyber attack on HHS Attempted to Slow Down Coronavirus Response." *US News & World Report*, March 16, 2020, *https://www.usnews.com/news/health-news/articles/2020-03-16/cyberattack-on-department-of-health-and-human-services-attempted-to-slow-down-coronavirus-response*.

Lawson P., et al. "Email Phishing and Signal Detection: How Persuasion Principles And Personality Influence Response Patterns and Accuracy." *Applied Ergonomics*, vol. 86, July 2020, doi:10.1016/j.apergo.2020.103084.

Leiber, N. "Using Drones to Make Peace, Not War." *Businessweek*. October 23, 2014: 54–56. Print.

Levine, B. "San Jose Police Discover That, Yes, They Did Buy a Drone." *ArsTechnica*. Condé Nast, July 30, 2014. Web. October 26, 2014, *https://venturebeat.com/2014/07/30/san-jose-police-discover-that-yes-they-did-buy-a-drone/*.

Link, D. "A Semi-Automated Content Moderation Workflow for Humanitarian Situation Assessments" in *Emergency and Disaster Management: Concepts, Methodologies, Tools, and Applications* (1-5225-6196-X, 978-1-5225-6196-5), (882).

Lyashenko, V., et al. "Tools for Investigating the Phishing Attacks Dynamics." 2018 International Scientific-Practical Conference Problems of Infocommunications. Science and Technology (PIC S&T), February 4, 2019, doi:10.1109/infocommst.2018.8632100.

Malatji, M., et al. "Validation of a Socio-Technical Management Process for Optimising Cybersecurity Practices." *Computers & Security*, vol. 95, 2020, p. 101846., doi:10.1016/j.cose.2020.101846.

Marks, J. (April 1, 2020). "The Cybersecurity 202: Coronavirus pandemic unleashes unprecedented number of online scams", *The Washington Post*, *https://www.washingtonpost.com/news/powerpost/paloma/the-cybersecurity-202/2020/04/01/the-cybersecurity-202-coronavirus-pandemic-unleashes-unprecedented-number-of-online-scams/5e83799b88e0fa101a757098/*.

Marzullo, K. "Report: SJPD Is First Local Agency to Get a Drone." *ABC7* San Francisco. ABC Inc., July 8, 2021. Web. October 26, 2014, *https://abc7news.com/san-jose-police-sjpd-drones-drone/229565/*.

McCabe, A. "The Fractured Statue Campaign: US Government Agency Targeted in Spear-Phishing Attacks." *Palo Alto Networks*, March 12, 2020, *https://unit42.paloaltonetworks.com/the-fractured-statue-*

campaign-u-s-government-targeted-in-spear-phishing-attacks/.

Moir, N. "ISIL Radicalization, Recruitment, and Social Media Operations in Indonesia, Malaysia, and the Philippines." *PRISM | National Defense University*, September 14, 2017, *https://cco.ndu.edu/News/Article/1299567/isil-radicalization-recruitment-and-social-media-operations-in-indonesia-malays/cco*.

Montero-Canela, R., et al. "Fractional Chaos Based-Cryptosystem for Generating Encryption Keys in Ad Hoc Networks," vol. 97, February 2020, p.102005. doi:10.1016 j.adhoc.2019.102005.

Morovati, K., & Kadam, S. S. (2019). "Detection of Phishing Emails with Email Forensic Analysis and Machine Learning Techniques." *International Journal of Cyber-Security and Digital Forensics*, 8 (2), 98+.

New Osterman Survey on the "Phishing Prevention Perception Gap Reveals Disconnect Between C-Suite and Cybersecurity Professionals." (March 10, 2020). *PR Newswire*, p. NA.

Nixon, R. "Social Media in Afghanistan Takes On Life of Its Own." *The New York Times*. April 29, 2014. Web. April 30, 2014.

Norris, G. and Brookes, A. "Personality, Emotion, and Individual Differences in Response to Online Fraud." *Personality and Individual Differences*, January 8, 2020, doi:10.1016/j.paid.2020.109847.

Parks, L. (09/2019). "Dirty Data: Content Moderation, Regulatory Outsourcing, and The Cleaners." *Film Quarterly*.

Proofpoint. "What is Email Spoofing?" *https://www.proofpoint.com/us/threat-reference/email-spoofing*.

Raja, E., and Ravi R. "A Performance Analysis of Software Defined Network Based Prevention on Phishing Attack in Cyberspace Using a Deep Machine Learning with CANTINA Approach (DMLCA)." *Computer Communications*, vol. 153, March 1, 2020, pp. 375–81, doi:10.1016/j.comcom.2019.11.047.

Reyes, R. "LAPD Drones Approved for Permanent Use." *Government Fleet*, September 19, 2019, *https://www.government-fleet.com/340648/lapd-drones-approved-for-permanent-use*.

Riedl, J. M, et al. "The Downsides of Digital Labor: Exploring the Toll Incivility Takes on Online Comment Moderators." *Computers in Human Behavior*, vol. 107, 2020, p. 106262, doi:10.1016/j.chb.2020.106262.

Ryan-Mosley, T., and Strong, J. "The Activist Dismantling Racist Police Algorithms." MIT Technology Review, June 5, 2020, *https://www.technologyreview.com/2020/06/05/1002709/the-activist-dismantling-racist-police-algorithms/*.

Salonga, R. "San Jose Police Drone Inflames Surveillance." *San Jose Mercury News*. August 1, 2014. Web. July 8, 2021, *https://www.mercurynews.com/2014/07/31/san-jose-police-drone-inflames-surveillance-state-rumblings/*.

Sanger, D. E., and Perlroth, N. "More Hacking Attacks Found as Officials Warn of 'Grave Risk' to U.S. Government." *The New York Times*, June 15, 2021, *https://www.nytimes.com/2020/12/17/us/politics/russia-cyber-hack-trump.html*.

San Jose Police Department, "SJPD Fallen Officers.", October 25, 1999. Web. December 4, 2014, *https://www.sjpd.org/about-us/inside-sjpd/department-information/fallen-officers*.

Schaake, M., and Reich, R. "Election 2020: Content Moderation and Accountability." Human-Centered Artificial Intelligence Cyber Policy Issue Brief, Stanford University, October 2020.

Smith, J. C. (01/01/2019). "Whaling" in Global Crime: An Encyclopedia of Cyber Theft, Weapons Sales, and Other Illegal Activities: M-Z (1-4408-6015-7, 978-1-4408-6015-7), (645).

Studies from Center for Democracy and Technology Have Provided New Information about Big Data (No amount of AI in content moderation will solve filtering's prior-restraint problem). (May 25, 2020). Bioterrorism Week, p. 689.

Teh Sen, J., et al. "Implementation and Practical Problems of Chaos-Based Cryptography Revisited." Journal of Information Security and Applications, vol. 50, February 2020, p. 102421, doi:10.1016/j.jisa.2019.102421.

"The Emotional Toll Of Content Moderation." (December 21, 2019). *Weekend All Things Considered,* *https://www.npr.org/2019/12/21/790492548/the-emotional-toll-of-content-moderation?t=1649424602990*.

Tschakert, F., and Ngamsuriyaroj, S. "Effectiveness of and User Preferences for Security Awareness Training Methodologies." *Heliyon*, vol. 5, no. 6, June 2019, doi:10.1016/j.heliyon.2019.e02010.

"Two Hat Is Changing the Landscape of Content Moderation With New Image Recognition Technology." (January 17, 2019). PR Newswire.

"Unmanned Aircraft Systems." Federal Aviation Administration. July 8, 2021. Web. October 24, 2014. *https://www.faa.gov/uas/*.

U.S. CONST. amend. IV., *https://www.law.cornell.edu/constitution/fourth_amendme nt*.

U.S. Department of Defense. "DOD Developing Small, Unmanned Aerial System for Warfighters", August 20, 2020. *https://www.defense.gov/News/News-Stories/Article/Article/2318919/dod-developing-small-unmanned-aerial-system-for-warfighters/*.

Vishwanath, A. "Spear Phishing Has Become Even More Dangerous." *CNN*, September 1, 2018, *https://edition.cnn.com/2018/09/01/opinions/spear-phishing-has-become-even-more-dangerous-opinion-vishwanath/index.html*.

Willon, P., and Mason, M. (2014) "Governor Vetoes Bill That Would Have Limited Police Use of Drones." *Los Angeles Times*, *https://www.latimes.com/local/political/la-me-ln-governor-vetoes-bill-to-limit-police-use-of-drones-20140928-story.html*.

Xie, J., et al. "A Network Covert Timing Channel Detection Method Based on Chaos Theory and Threshold Secret Sharing." 2020 IEEE 4th Information Technology, Networking, Electronic and Automation Control Conference (ITNEC), June 12, 2020, doi:10.1109 itnec48623.2020.9085024.

Yang, F., et al. "Sequence of Routes to Chaos in a Lorenz-Type System." *Discrete Dynamics in Nature and Society*, vol. 2020, 2020, pp. 1–10, doi:10.1155/2020/3162170.

Zakrzewski, C. "The Technology 202: Lawmakers plan to ratchet up pressure on tech companies' content moderation practices." (April 9, 2019). *The Washington Post*, *https://www.washingtonpost.com/news/powerpost/paloma /the-cybersecurity-202/2020/04/01/the-cybersecurity-202- coronavirus-pandemic-unleashes-unprecedented-number- of-online-scams/5e83799b88e0fa101a757098/*.

Zakrzewski, C. "The Technology 202: NYU report calls social media titans to stop outsourcing content moderation." (June 8, 2020). *The Washington Post*, *https://www.washingtonpost.com/news/powerpost/paloma /the-technology-202/2020/06/08/the-technology-202-nyu- report-calls-social-media-titans-to-stop-outsourcing- content-moderation/5edd3806602ff12947e865d2/*.

Zimmermann, V. Moving from a "Human-as-Problem" to a "Human-as-Solution" Cybersecurity Mindset. *International Journal of Human-Computer Studies*, vol. 131, 2019, pp. 169–87, doi:10.1016/j.ijhcs.2019.05.005.

FURTHER READING

IT Governance Publishing (ITGP) is the world's leading publisher for governance and compliance. Our industry-leading pocket guides, books, training resources, and toolkits are written by real-world practitioners and thought leaders. They are used globally by audiences of all levels, from students to C-suite executives.

Our high-quality publications cover all IT governance, risk, and compliance frameworks, and are available in a range of formats. This ensures our customers can access the information they need in the way they need it.

Other publications you may find useful include:

- *The Cyber Security Handbook – Prepare for, respond to and recover from cyber attacks* by Alan Calder, *https://www.itgovernancepublishing.co.uk/product/the-cyber-security-handbook-prepare-for-respond-to-and-recover-from-cyber-attacks*
- *The Ransomware Threat Landscape – Prepare for, recognise and survive ransomware attacks* by Alan Calder, *https://www.itgovernancepublishing.co.uk/product/the-ransomware-threat-landscape*
- *Cyberwar, Cyberterror, Cybercrime & Cyberactivism – An in-depth guide to the role of standards in the cybersecurity environment, Second edition* by Julie Mehan,

*https://www.itgovernancepublishing.co.uk/product/cy
berwar-cyberterror-cybercrime-cyberactivism-2nd-
edition*

For more information on ITGP and branded publishing services, and to view our full list of publications, visit *https://www.itgovernancepublishing.co.uk/*.

To receive regular updates from ITGP, including information on new publications in your area(s) of interest, sign up for our newsletter at *https://www.itgovernancepublishing.co.uk/topic/newslette
r*.

Branded publishing

Through our branded publishing service, you can customize ITGP publications with your organization's branding.

Find out more at:
*https://www.itgovernancepublishing.co.uk/topic/branded-
publishing-services*.

Related services

ITGP is part of GRC International Group, which offers a comprehensive range of complementary products and services to help organizations meet their objectives.

For a full range of resources on cyber security visit *https://www.itgovernanceusa.com/cybersecurity-
solutions*.

Training services

The IT Governance training program is built on our extensive practical experience designing and implementing management systems based on ISO standards, best practice, and regulations.

Our courses help attendees develop practical skills and comply with contractual and regulatory requirements. They also support career development via recognized qualifications.

Learn more about our training courses in cyber security and view the full course catalog at *https://www.itgovernanceusa.com/training*.

Professional services and consultancy

We are a leading global consultancy of IT governance, risk management, and compliance solutions. We advise organizations around the world on their most critical issues, and present cost-saving and risk-reducing solutions based on international best practice and frameworks.

We offer a wide range of delivery methods to suit all budgets, timescales, and preferred project approaches.

Find out how our consultancy services can help your organization at

https://www.itgovernanceusa.com/consulting.

Industry news

Want to stay up to date with the latest developments and resources in the IT governance and compliance market? Subscribe to our weekly round-up newsletter and we will send you mobile-friendly emails with fresh news and features about your preferred areas of interest, as well as unmissable offers and free resources to help you successfully start your projects. *https://www.itgovernanceusa.com/weekly-round-up*.

EU for product safety is Stephen Evans, The Mill Enterprise Hub, Stagreenan, Drogheda, Co. Louth, A92 CD3D, Ireland. (servicecentre@itgovernance.eu)

www.ingramcontent.com/pod-product-compliance
Lightning Source LLC
Chambersburg PA
CBHW061830220326
41599CB00027B/5251